MW01489157

Box Turtles, Hooligans, and Love, Sweet Love

Mary Dansak

Library of Congress Control Number: 2025900002

Author photo by Lizzie Hudson.

First printing, January 2025

Published by Little Green Notebook
Auburn, AL 36830

littlegreennotebookpublishing@gmail.com

ISBN: 9798304546874

Table of Contents

Dedication

This book is dedicated to the beautiful state of Alabama, to her rivers and rocks, her critters and trees, her mountains, her plains, and her sandy beaches, and to all the humans who work to preserve her rich biodiversity.

Introduction

This book is a collection of the columns I wrote between August 2022 and December 2023. After much gnashing of teeth over the arrangement of this book, I decided to present the columns in the order they were written, providing the date they were published in *The Auburn Villager* for each. I've included most of the columns I wrote during this time, omitting a scant few that were either too timely or too local for universal appeal.

It is a joy to write a weekly column in which I am able to share my profound love of the natural world, to reflect on experiences I think might resonate with others, and to muse upon wonders.

I hope you enjoy reading my odes, rambles, musings, and reflections as much as I enjoy writing them.

The Orb Weavers of August
August 4, 2022

It's summertime in Alabama. Though the heat is unbearable, summer brings its own treasures to our state. There's nothing like a morning cicada serenade with your first cup of coffee. I was doing just that, sipping coffee and admiring the gentle stirring of the morning breeze through our Tibetan prayer flags, when I realized there was no morning breeze, and only one of the flags was moving. My eyes followed a thin golden strand from the flag to an empty space between two trees and there she was, a magnificent *Trichonephila clavipes*, otherwise known as a golden orb weaver.

Our yard, and likely yours, is loaded with these enormous orb weavers. They're often confused with black and yellow garden spiders, *Argiope aurantia*, which spin a large zigzag called a stabilimentum into their webs. These lovelies are nicknamed "writing spiders," which is fabulous for many reasons, but for now I'll call them "garden spiders."

Both of these gigantic spiders are orb weavers, referring to the large, circular webs they spin.

They exhibit sexual dimorphism, meaning the males and females have substantially different bodies. In fact, I was talking baby talk recently to a male golden orb weaver thinking he was a baby. It turns out the males are just small. The females are much larger. If you were to put a female golden orb weaver on your face it might cover your entire cheek from jaw to eye. If you choose to measure this way, please send me a photograph.

Most spiders have two claws on their front feet for web building; our orb weaving friends have an extra claw. It's no wonder their webs are spectacular. Speaking of webs, I've been plagued with guilt because, as a child, I used to purposefully knock down our garden spider's web so I could watch her build it back. It turns out that they routinely knock down, eat, and rebuild their webs, sometimes daily, so now I don't feel so bad. My friend Padric claims he trains his orb weavers by walking through the webs a few times until the spiders learn to build them higher, and now I have a new morning project.

Right now, I'm not seeing garden spiders, but oh my stars, the golden orb weavers are here in abundance! According to my notebooks, they're three weeks earlier than last year. It turns out they thrive in our increasingly warming urban heat islands, a golden lining to an unfortunate reality.

These orb weavers live for about a year, hiding out for most of their lives until late summer when it's time for reproduction. Unlike many of her friends, a female golden orb weaver will not eat her mate, but continue to cohabitate with him after their business is done. Once females lay their egg sacs, they live on for a about a month while the male orb weavers hold on for just two weeks.

I'd be remiss not to mention the superpowers of the golden orb weavers' webs. Stronger than Kevlar, they've been used by a couple of avant-garde designers to make a shawl and cape. Some fishermen on the Indo-Pacific coasts ball up the webs and toss them into the ocean where they blossom into bait fish nets. My favorite human use for this spider silk is its role in mammalian neurosurgery where it encourages neuronal growth, its effectiveness compounded with strength by eluding the human immune system.

I like to ponder my immune system accepting spider silk when I'm feeling sad.

To identify which orb weaving spider you are seeing when sipping your morning coffee, take note of the abdomen. Garden spiders have oval-shaped abdomens, with varying black and yellow color patterns. The golden orb weavers' abdomens are a squared-off cylindrical shape and are yellow with white dots. Next check the webs. As mentioned, the garden spiders' webs sport the tell-tale zipper-like stabilimentum, a dependable identifier. Finally, Golden orb weavers have extra hairy parts on their legs, like bottle brushes. Neither species is harmful to anything but their prey which includes mosquitoes. Have you thanked an orb weaver today?

I suppose it doesn't matter much if you know which orb weaver is which, but humans have demonstrated repeatedly that to name something is to love it, and so I encourage everyone to learn the names of their backyard critters.

Postscript: After writing this column, I anxiously awaited the return of the golden orb weavers the next August. Alas, I didn't see a single one. We had an unusually cold December that year, and some speculate that perhaps that deep freeze killed the spiderlings in their egg case. Another August has since come and gone and they still have not returned. I await them, ever hopeful every August.

Possums On My Mind
August 11, 2022

Possums are trendy these days. They show up all over my social media. My daughter sports a possum shoulder bag (okay, I gave it to her, but still). Last night my husband Joe deterred a possum from crossing the street, guiding it back to the safety of our yard. I even have a possum mood bracelet. It's no wonder I've got possums on my mind.

I'm all emotional when it comes to possums. I think you'll understand after hearing about my childhood possum friend.

I was in 5th grade when my dad came home with a baby possum, *Didelphis virginiana*. He'd found her on the side of the road after stopping to check on an adult possum who'd been hit by a car. The adult was dead, but there were two babies with her, alive and clinging to her fur. He took one to a friend and brought one home.

Our new charge was about two months old. If you do not know what a two-month-old baby possum looks like, please put this down and go find a picture at once. They are exquisite creatures. Their ears are like delicate shells, always perched forward in curiosity. Their beady black eyes are bright with questions. And oh, those twitching whiskers and that pale pink nose! Be still my heart!

"She'll live in your hair a while," my dad said casually, introducing the possum to my head. She crawled into my tangled hair and held on.

When possums are born, they are no bigger than a bean. Although they can give birth to up to 25 babies, only the lucky ones that find the mama's pouch and latch onto a teat survive. After about two months, they leave the pouch and cling to their mama's fur, visiting the pouch until weaned. Around four months, they let go of their safety net. Young possums will stay with their mamas for a year, but if one should straggle, or fall off the possum-bus early, they are on their own.

For several weeks I played mama possum. The baby ate in my hair. She slept in my hair. And yes, we had to wash my hair, possum and all, frequently. When she grew brave enough to leave the safety of my frazzled locks, I got a haircut. My own mama took a picture of me standing in the azaleas wearing a hot pink dress, my newly bobbed hair gleaming, and the wee, liberated possum perched on my shoulder. I was as proud a mama possum as any mama possum ever was.

We named our pet possum Rock Hudson, then realizing she was female, we changed her name to Rockaposs. Rockaposs grew into a ridiculously fat adult. I brushed her and cuddled with her daily. I dressed her in doll clothes and put ribbons in her silky hair.

On special occasions, she treated us to a "possumpoo." We would sit on our couch and put Rockaposs on the cushions behind us. "Choose me," I'd will with all my might while my brother willed the same. Oh, the thrill when she chose me, gently placing her paws upon the crown of my head. Millimeter by millimeter, she'd lean forward and salivate onto my hair, then massage her saliva into my hair, massaging my scalp with her dexterous front paws. Eventually my whole head would be dripping wet, and to top it off, Robert would be exceedingly jealous.

At least she took turns.

I understand if you are feeling grossed-out. Now imagine the honor. Even in its most perplexing manifestations, the affection of a wild thing is precious, reverent, holy. Okay, maybe getting slimed by an obese possum is not holy, but it is something.

Eventually the time came to release our beloved possum. We took her to a friend's house who not only lived in the woods but had raised one of Rockaposs's siblings. They promised to leave food out for her as long as she came around. My dad tried to console me that this was for the best. Robert and I fought over who would be the last one to touch her. Finally, we both laid a hand on her, my parents counted to three, and my dad chased us to the car. We stared at Rockaposs through the back window as we pulled away.

I cried and cried when we released Rockaposs. Caring for and returning animals into the wild was hard, but it taught me to respect the true nature of wildlife.

An Unlikely Celebration
August 18, 2022

Dust off the box fans and grab the OFF! It's almost time to celebrate World Mosquito Day! August 20, 2022, marks the 125th anniversary of the day Sir Ronald Ross discovered that mosquitoes carry malaria, thus changing the course of history.

Despite my love of creepy-crawly-critters and all things found in my backyard, mosquitoes are an exception. Let's explore how we can enjoy our celebration without including the celebrants.

Shall we spray the yard? Although pest control companies have regulations about spraying within range of flower beds and taking measures to control drift, this is not an option for us. The thigh-high weeds and thick vegetation in our wooded yard provide protection for insect larvae year-round, and we welcome the throngs of lightning bugs, cicadas, katydids, spiders, toads, turtles, snakes, and other creepy-crawlies who take shelter here.

Perhaps we should buy a bug zapper. They lure mosquitoes in with a pretty blue light then electrocute them. But you know what mosquitoes find more attractive than the light? Your breath, your glow, your essence. Mosquitoes find us by the carbon dioxide we expel and will quickly detour from the lights to our tender skin. Entomologists who have studied zappers conclude that biting insects represent a tiny portion of insects they kill, and those that are zapped are necessary for a thriving ecosystem. No bug zapper for us.

Bats are a lovable option for mosquito control. While the claim that bats eat 1000 mosquitoes a night is an unsubstantiated exaggeration, bats do drive mosquito populations down. For the longest time we had bats nesting in our canister lights outside. We removed the lightbulbs to make more room for the bats and enjoyed visiting with them frequently. Now they're gone, alas, so I guess it's time to put up a bat house.

Speaking of bats, our friend Vince hung two bat houses in his yard. He painted the words, "Welcome Bats" on one, and nothing on the other. The bats readily moved into the "Welcome Bats" house, but the unadorned house remains empty. We hypothesize that Vince's bats are literate, but further research is necessary.

One promising mosquito control technique involves genetic modification. In March 2022, the FDA approved the continuation of trials in California and Florida for testing the effects of releasing genetically modified male mosquitoes whose female offspring cannot survive. The first round of trials has been highly effective at reducing the numbers of a particular mosquito species which carries yellow fever, dengue, and Zika virus. I realize that many people are wary of GMOs. I am not, based on my understanding of how DNA breaks down into nucleotides that end up in the nucleotide soup in my cells, but I am far from the expert. Regardless, this testing is not happening in my home state.

What to do? As for me and my house, we rely on Deep Woods OFF! and fans. Fans combat mosquitoes in two ways. They scatter your carbon dioxide, making you less of a target, and they create a hostile environment. Mosquitoes cannot fly in winds over 10 mph, so a nice box fan on the porch or in the yard is an excellent deterrent, and you get a cool breeze as a summer bonus.

After repelling our mosquitoes safely and responsibly, let's raise a glass to them. Mosquitoes are important pollinators; in fact, there is a species of orchid that depends on them entirely. Additionally, mosquitoes and their larvae are part of the food chain. Mess with those links and ecosystems become unstable.

Finally, let's all be amazed at the mosquito's proboscis. It is slightly softer at the tip, with smaller serrated blades increasing in size further up. This design eliminates pain. The elegant proboscis has two channels: one to deliver mosquito saliva to the skin which acts as a numbing agent (and later, causes itchiness), and another to slurp up the blood. Bioengineers have shared this design with syringe manufacturers, making our hypodermic experiences less painful. Be on the lookout for double-barreled micro-needles coming your way!

Despite these marvels, it'll be a cold day in an Alabama summer before I fall in love with mosquitoes. Still, as with all creatures great and small, I enjoy learning their secrets. As you battle mosquitoes this summer, please keep in mind the other flying friends taking refuge in your yard.

Foxes in Our Midst
August 25, 2022

Back in my teenage days, I once came home and found an angry, frightened fox huddled on a blanket in the corner of the kitchen.

"Hey there," I whispered. He pressed against the wall and stared at me with huge, black eyes as he snarled and bared his teeth.

"That's Sid Vicious," my dad said. I was taken aback, more surprised that my dad knew the name of the bass player for the Sex Pistols than I was at finding a fox in the kitchen.

Sid Vicious was injured and needed a place to heal. After a week or so, he went back into the wild and I never saw him again.

For the longest time, I never saw any foxes. Then about 10 years ago, according to my trusty notebook, we began seeing a fox now and then in our neighborhood. And just a few days ago, my husband Joe saw two foxes in a highly congested area of town.

As they are notoriously wary of humans, it's some powerful luck to see a fox. Sometimes I'll see a red fox with some black and white highlights. Other times I'll see a fox that is colored like a coyote. I'll share some information with you I found in clearing up my own confusion about what foxes look like.

Here in southeast, we see gray foxes (*Urocyon cinereoargenteus*) and red foxes *(Vulpes vulpes)* in our suburban areas. Gray foxes are usually gray with a lighter underside and a black stripe down their backs ending in a black-tipped tail. Red foxes usually sport the story-book red coat, black legs, and a white-tipped tail. Colors vary in both species.

And what about coyotes? Coyotes are larger than foxes, with longer legs, ears, and snouts. Coyotes' faces are more dog-like, whereas foxes have almost cat-like faces.

If you're confused, don't fret. These three canids routinely stump citizen scientists. Speaking of coyotes and foxes, we need to dispel a myth. There are no coyfoxes; coyotes and foxes are in different genera and do not share the same number of chromosomes. They are not able to interbreed. The same is true with dogs and foxes. Simply put, foxes stick to foxes and that's that.

Should we fret over foxes in our midst? It never occurred to me to fear for our dogs when it comes to foxes, and it turns out I'm right. Foxes are not interested in animals that large. They are, however, a threat to small animals like rats, rabbits, and chipmunks. Our friends with chickens will be wise to protect them in fox-proof cages (chicken wire does not cut it, I hear). Foxes are generally not a nuisance, mind the occasional tipped garbage can or stolen pet food. And we don't need to worry about the neighborhood going to the foxes. They are territorial and will limit their own populations. As for rabies, instances in foxes are extremely rare.

So, foxes are beautiful, not a nuisance, and no threat to my family. Maybe I should get a pet fox. Like all sound-minded children, I dreamed of having a pet fox. Like some sound-minded adults, I still carry this dream in my heart.

Perhaps you've heard of Dmitri Belyaev's experiment in Russia back in the 1950s. He set out to discover the effects of selectively breeding the least wild-acting silver foxes in a set, repeating with each generation. After ten generations he had a slew of silver foxes who eagerly sought out human companionship, behaving a lot like dogs. Are these foxes truly domesticated, meaning their genes have permanently mutated to accommodate life with humans? Questions abound.

Warning: Do not look up pictures of domesticated silver foxes. Your heart will burst from the cuteness.

Did you look? Now you want one, right? I believe you can purchase a domesticated fox from Russa today for about $10,000. But would it be legal? As of this writing, 16 states allow people to own foxes with varying degrees of regulation. There seems to be no rhyme or reason as to which states allow which species. In Alabama, it is not legal to own a fox for any reason. There goes my fox fantasy. They seem like a handful to be honest, and not altogether happy as pets.

Here is a bit more about your new canine neighbors. Male foxes are called dogs, females are called vixens, and babies are called kits. A family of foxes is a skulk or a leash of foxes. All foxes have exceptionally keen hearing, and they can make up to 40 different sounds with gekkering, a throaty chatter, being the most common. Finally, don't feed the foxes. This will only make life harder for them in the long run.

If you spot a fox in your neighborhood, no action is required. Just bask in the beauty of this exquisite creature and if necessary, batten down the hatches on the chicken coop.

The Mona Lisa of Snakes
September 1, 2022

I must have seen five gray rat snakes this week. That's good; they are one of my favorites. It's common knowledge that these snakes are beneficial. They keep the rodent populations down and pose no threat to humans or our pets, except maybe the chicken owners (again) who might lose a few eggs to these opportunistic carnivores. Regardless, folks have negative feelings about all snakes. To know me is to love me, they say, so let's get to know one of my favorite serpents.

What exactly is a rat snake? It's complicated. You know those herpetologists, always embroiled in controversy over this and that. There are over 45 species of rat snakes who were all comfortably categorized in various taxa, or taxonomic groups, until 2002 when along came new DNA analysis of the North American rat snakes. This caused a great brouhaha among herpetologists. They shook their fists and rattled their whiskey glasses while grumbling over how to make sense of these findings. I think the science is settled now, placing all the North American rat snakes in the genus *Pantherophis,* "panther snake," which includes eight living species.

Here in the southeast USA, we have two species among us: gray rat snakes and corn snakes. While corn snakes are more popular due to their spectacular colors as well as easy-going disposition, my favorite is the good old gray rat snake.

Let's say you encounter a snake. Here's a simple identification key. Is it climbing a tree, house, or other surface that seems unclimbable? It's likely a gray rat snake. Excellent climbers, rat snakes will even nest in trees, though they are primarily terrestrial burrowers. Is it startlingly long? Though typically three to five feet in length, gray rats can grow to seven feet, one of the longest North American snakes. Is it gray with black rectangles ("saddles") on its back? Congratulations! It's a gray rat snake.

When threatened, gray rat snakes go stock-still. Sometimes they kink their bodies in a zig-zag, mimicking a stick. We saw this in action not long ago when my husband rescued a motionless gray rat snake from the road. It was so still that Joe thought it had been hit by a car. We were both surprised to see its body contorted and kinked. All for show, we learned, and soon the rat snake was living large in our wooded yard.

If you come upon a gray rat snake, you may be lucky enough to get a good look at its face. Given their defensive stillness, odds are in your favor. Oh my stars, they're so cute! Their eyes are round as saucers, their expressions calm with an ever so slight look of amusement. You might say they are the Mona Lisas of snakes. Unlike pit vipers, who look like they want to scare you, rat snakes seem to want to calm you, to help you make peace with the drudgeries of your daily life. I find that staring into the eyes of a gray rat snake elevates my oxytocin levels. (Warning: Do not confuse the genera *Pantherophis* and *Panthera*. Should you encounter a leopard, *Panthera pardus*, do NOT look it in the eye. It will attack.)

As for pets, gray rats are excellent starter serpents. They are generally easy to keep in captivity. I've had several in my time. I even took one, Scarlett to class with me when I was in college.

I used to leave Scarlett to roam my tiny college duplex, a practice I do NOT endorse. Sometimes, she would wrap herself up in the coils in my sofa. There was no coaxing her out of the tangle. Believe it or not, I'd just leave her, and she'd return herself to her tank on her own time. On the day in question, I assumed she was nestled in her sofa coils when I grabbed my backpack and ran off to class. Imagine my amusement when I reached into my bag to retrieve my chemistry notes and lo, retrieved a snake instead. Imagine the other students' reactions when I held her up for all to see.

Before leaving this rat snake revelry, I must recognize my father, Robert H. Mount, who spoke Parseltongue long before J.K.

Rowling coined the term. (For those unfamiliar with the Harry Potter books, Parseltongue is the ability to communicate with snakes.) The cobbler's children go barefooted they say, and my dad did not come home from work every day and fill our heads with facts and figures about the critters he worked with. Instead, he made these slithering beasts a part of our everyday life.

It thrills me to no end that I come upon snakes frequently, calmly keeping the peaceable kingdom balanced.

To learn more about our snakes, see the book, *Lizards and Snakes of Alabama,* by Craig Guyar, Mark A. Bailey, and Robert H. Mount, University of Alabama Press. Unbiased side-note: this colorful book makes an excellent gift for the herpers and herper wanna-bes in your life.

A Day on a River
September 15, 2022

We finally tossed my dad's ashes into the river. I debated admitting this publicly, but after checking the books I understand that we didn't break any laws, so here we are.

Family and friends gathered at Horseshoe Bend where the rock piers of the Miller Covered Bridge, now vanished but for these old supports, provide a stately entrance to a promising journey. Within seconds our daughter Emma capsized in her kayak. I shared a canoe with my uncle Richard, and after a ridiculous show of going nowhere, our friend Joseph hollered, "What are y'all doing?" We were backwards in our seats. Once we all got facing the right way and settled into our respective boats, two bald eagles swooped down from the trees. They turned and led us onward, our guides for this important mission.

Every time I'm on a river I ask myself why I don't spend more time on rivers. As a child my very favorite days were those when my mom and dad, brother, dog, and for a time pet crow would load up in the Volkswagen bug and head west to spend the day on Mussel Creek. After what seemed like a long car ride (in reality about 30 minutes) we'd park and walk down a brambly path to the creek, then unpack our bags and make our little home for the day.

Robert and I collected firewood while Mama and Daddy set up the bonfire and spread towels on the ground. We were not one of those fancy families that brought chairs to the creek. Duties done, we'd go our separate ways. Mama would alternate between walking slowly in the shallow water collecting rocks and sitting on the creekbank writing in her journal. Robert and Daddy would head upstream to look for snakes and turtles, particularly soft-shelled turtles, which I adored. They looked like our friend Nancy Lyle in the face.

I followed Daddy and Robert at a slow pace. I'd pull myself along in the water with my arms, staying close to the creekbank. I collected clay to make little pots, and painted my face yellow and red with what I called "Indian War Paint" before I knew better. Sometimes I'd find a spring and drink straight from the clear trickle that emerged from the mud.

Later we'd build a fire and roast hot dogs. I no longer eat hotdogs, but if I did, I'd want them on cheap white buns with mustard, ketchup, and a generous splash of OFF! Ah, the taste of childhood! For dessert we'd roast marshmallows. I roasted my marshmallows to a fine golden brown; Robert put his right into the flames of the bonfire and blackened them. The horror.

I never wanted to leave. Aside from loving the creek itself, the ride home was miserable. There was no way to rinse and shake away the sand. It dug into our legs which were pressed against the plastic seats. Robert and I pushed our dog back and forth, trying desperately to get just a few inches further from his wet-dog stench. Bowzer thought it was fun and wagged his tail.

When my parents announced their divorce, I realized at once there'd be no more trips to Mussel Creek and the anguish was fire in my chest. I was right, of course, and now I would not even be able to find my way to the spot where we spent so many glorious Sundays.

I doubt anyone could find it; there's no such place as Mussel Creek. My dad nicknamed it that for the bounty of freshwater mussels there. Our Mussel Creek is actually the Uphapee Creek, a tributary of the Tallapoosa, the big, beautiful river where we finally dumped his ashes.

All this came back to me as we paddled along, looking for the proper spot to toss my dad to the river. We finally gathered beside a glorious spray of wildflowers – yellow asters, purple ironweed, scarlet lobelia – and banded our boats together. We may or may not have toasted my dad with his favorite whisky, "Evil Willie." My stepmother Janie announced the return of Bob to the environment which he fought his whole life to protect, Joseph lit the sage, and we each tossed a share of Bob's ashes into the water. Someone hollered, "Hip hip hooray!" High and holy church indeed.

Off he goes now to meet up with the Alabama River, the Gulf of Mexico, and so on.

I could get nice and sciencey now and talk about the fact that Alabama has more navigable rivers than any other state, or marvel the wonders of freshwater mussels with you, but today is a day for sentiment, for remembering that those rivers run through me, through you, through my family and friends, through eagles and mussels and turtles and salamanders, and to remind myself that of all the days, days on a river are best.

Hope in a Hummingbird
September 22, 2022

Mid-September mornings in Alabama bring a tease of cooler temperatures, though the month repeatedly disappoints with insufferable heat as soon as the sun crests over the trees. Still, here I am, sitting on the porch with coffee and dogs, loving the chill on my bare arms. I welcome winter, unlike our migratory hummingbirds.

The hummingbirds are here with me now, divebombing each other for a seat at one of the two feeders which between them can host ten hummingbirds at a time but in reality only host one. As I write these words, two ruby throated hummingbirds are buzzing each other like the warriors they are. To them the feeder is just one flower, a tiny bit of nectar. This endless bounty of fuel is inconceivable to their brains. 22 million years of instructions to gather nectar from over 1000 flowers a day can't be undone this quickly.

The hummingbirds are fueling up for their journey south. They have a long way to go, between 500-2000 miles depending on which route they take, the longer route providing more places to stop for fuel. They will travel alone, facing predators (especially cats), windows, electric lights that never dim, heavy rains, heat, drought, and more. It appears that our backyard males have already headed out; only females are here this morning.

Most years we take the feeders down, following the rule of thumb to wait two weeks from the last siting. I depend on my little notebook for so much more than grocery lists: "Hummingbirds still here." "Nighthawks, bats, and dragonflies at dusk." "Laundry detergent. Bananas. Wine." "Binge-watching Better Call Saul."

I'm doing something different this year. I will keep my feeders up year-round for the old, the injured, and the weak. I've checked my sources and am convinced that my doing so will not disrupt migrations and may be a true respite for some little friends. If you choose to join me, remember to continue to change the food weekly and keep the feeders stocked. Starting winter feedings and then stopping is apparently more devastating than removing the feeders all together.

I think most folks will agree that hummingbirds astound, delight, and mystify. I recently learned about a hummingbird species in Tierra del Fuego and its surprising role in a mutualistic three-way relationship. Oh, my stars, get ready. This one is rich.

Down in South America at the end of the land there exists a flower, a wee marsupial, and a hummingbird who have coevolved in the most delightful way. Let's start with quintral (*Tristerix corymbosus*), a type of mistletoe that blooms with a splashy red blossom. It is the only source of nectar for pollinators in the winter. Enter the green-backed fire crown hummingbird (*Sephanoides sephanoides*), the only species of hummingbird in the temperate rainforests of southern South America and a vital pollinator to all the woody plants there. Quintral and green-backed fire crowns have a classic mutualistic relationship.

So how does the third party, the wee marsupial, fit in? In order for quintral seeds, the fruit of the hummingbird-flower relationship, to germinate, they must first pass through the digestive tract of this tiny mammal, the Monito del monte (*Dromiciops giroides*), the little monkey of the mountains. Despite its name and the fact that it has opposable thumbs and a prehensile tail, the Monito del monte is not a monkey at all. This tiny mouse-like critter eats the quintral seeds, scurries along the branches of trees, and deposits the now-viable seeds along its path.

Monitos del monte are the sole dispersers of quintral seeds. Only their guts can do the proper job. How precise nature can be. How perfect, how poetic.

Quintral is not only a source of food for green-backed fire crown hummingbirds, hundreds of families of birds and mammals rely on it for food and nesting material. This earns each member of our threesome the distinction of being a keystone species.

As their name suggests, Monitos del monte are extremely cute. Do yourself the favor of finding pictures. While searching, you may come across research that shows that their numbers are declining due to humans' destructive influences, mainly the introduction of domestic cats and deforestation. This may lead you to disturbing dreams like the one I had recently about migrating salmon, which resulted in my sitting upright in the bed and stammering, "We just don't have the right."

The forces that drive humans to destroy the very things that sustain us cannot be stopped by me keeping our hummingbird feeders clean and replenished, but they can make the difference in one hummingbird's life. And so, day after day, decision by decision, I take these tiny steps. Environmentalism depends on optimism and hope.

Hummingbirds have an excellent memory for places and plants that provide their food. I look forward to cooling temperatures soon, and hopefully await the return of our flittering friends with the dawn of spring.

Box Turtles, Hooligans, and Love Sweet Love
September 29, 2022

Like all self-respecting children, my brother Robert and I tried to dig a hole to China in our backyard. We didn't get there, but we did dig an impressive pit. After wearing blisters on our hands, we leaned back on our shovels and changed our plans. We decided to build a tipi over our pit, fashioning it with branches bound with baling twine and covered in pine straw. We attached a fake cow-skin flap over a hole for the door. Satisfied, we'd sit in the tipi, feet dangling in the pit. How fashionable, this mid-century modern tipi with its sunken den, was.

Oh, it was the grandest of forts, the Taj Mahal of neighborhood hideaways, until my hooligan brother's hooligan friends got mad at him and set it on fire.

All that remained was the pit, which we turned into a turtle pen. In no time we introduced our first turtle, then another. Robert and I, a passel of friends, and our dog Bowzer would hang out in the turtle pen with the "turteels," as we called them.

I lived a charmed childhood when it comes to critters and had no choice but to develop a profound love for nature, nurtured by long, uninterrupted hours of bonding with wild things and places. I am grateful for the times. These days such a turtle pen would be illegal. Due to declining numbers, box turtles are classified as protected by the Alabama Department of Conservation and Natural Resources. Removing a box turtle from the wild is illegal, as is keeping one as a pet.

Despite their declining numbers, I still come across box turtles from time to time, often in our woodsy yard. I pick them up and stare into their eyes, searching for secrets. If you are lucky enough to find one, I suggest doing the same. Here are some box turtle facts that will help you connect even more deeply with our terrapin friends.

In Alabama, we have two species of box turtles: Eastern Box Turtles (*Terrapine carolina*) and Florida Box Turtles (*Terrapine bauri*). As their name suggests, Florida Box Turtles live in the most southern counties bordering Florida. Eastern Box Turtles are found throughout the Southeast. Box turtles are named for their ability to close their shells, forming a box. The bottom part of the shell, the plastern, is hinged, enabling this feat.

When you stare into a box turtle's eyes, you'll notice some have red eyes, some have brown. Red eyes often indicate a male, but not always. The surest way to identify the sex of a box turtle is to check its plastern, the underside of its shell, for an indention in the hind area. The carapace, or top part of the shell, is often flared outward in that same area. These are both signs of a male, adaptations that are all the better for mating.

The sex of a box turtle is determined by the temperature of the nest in which the egg develops. Warmer nests result in female box turtles; cooler nests give rise to males.

One of the greatest threats to turtles is cars. And yes, you should make every effort to assist a turtle crossing a street! Please move them to the side of the road they are attempting to reach, far away from danger. Do not relocate them. Box turtles prefer to live their entire lives, which can span 50 years or more, in the same general vicinity.

Box turtles, like our hummingbird friends from last week, are on the move, but they aren't headed south. They're plumping up for a long winter's nap. Soon they'll crawl into a burrow and brumate, the reptile version of hibernating. They often return to the same hidey-hole, or hibernaculum, year after year. And for the word "hibernaculum," you're welcome.

We found two box turtles this week. One we helped up and over the street curb. The other my husband Joe brought in from the road in front of our house to show me. He put it on the table and it scurried to the other end. Speedy fellow. "Turteels! Turteels! Turteels on wheels!" I exclaimed, a remnant from childhood.

Who knew that this silly exclamation would change the course of our lives? A dozen years ago, when our daughter Sarah began dating the nephew of one of those childhood turtle-pen-hooligan friends, they found a turtle. Both exclaimed, simultaneously, "Turteels! Turteels! Turteels on wheels!" and with that, the lovey-dovey deal was sealed. Only fate could pull such a punch. Now they have two turtle-loving little girls of their own.

As for that speedy box turtle, we released it in our yard beside the tipi our friend Joseph lives in. With tipis in the yard and box turtles on the move, life keeps going around in beautiful circles.

Romancing the Gulf of Mexico
October 6, 2022

Despite an unorthodox childhood, our family had one traditional pastime: the summer vacation to the beach. We stayed at "Orange Beach Cottages," where pastel cottages named after flowers nestled on a rustic spit of land. Tall loblolly pines, old and craggly, provided shade leading up to the bay. We played at the land's edge for long hours swatting mosquitoes, chasing minnows, and paddling in the shallow water.

When I moved back to Alabama in adulthood I described the sugar white sands of the Gulf beaches to my husband Joe, who grew up in Massachusetts on the rugged, rocky shores of the Atlantic. He did not believe me. "Let's go," I said, and we loaded our three little girls into the car and off we went.

We stayed at the same Orange Beach Cottages. Nothing had changed. The girls played on the ramshackle pier, we ate hotdogs on picnic tables under the same tall pines and tossed and turned at night on thin mattresses, the metal coils poking our sunburned backs. We took the girls to the proper beach at the state park once a day but spent most of our time at the bay, sheltered from the glaring sun.

The Orange Beach Cottages were razed just months after our visit. We continued our family trips to Orange Beach, but now we stayed in high-rise condos. While nothing compares to the sugar white sands and glorious blues of the Gulf, my childhood beach was gone. Only the Gulf State Park Pier remained as a testament. Ever spooky, ever evocative, ever brimming with buzzing lights, a blue-billion bugs hovering round, and mumbled conversations from fisherfolk, the pier stood steadfastly on, unchanged since my childhood. We visited at night, avoiding the daytime sun.

One year a hurricane foiled our plans and we decided to try a new beach vacation. Somewhat randomly, we found a beach in South Carolina with no high-rise condos, only one grocery store, and acres and acres of protected wetlands. We have returned to this spot annually for the last 18 years.

My first venture back to the Gulf was to a conference in 2013. Alone, after the PowerPoints were saved and the computers shut down, I retreated to the Gulf State Park and dug my feet into memories of the wild beach of my childhood. I visited the pier and I was overwhelmed with a deep melancholy.

Since then, I've returned to the Gulf for more conferences, and lately a few girls' trips with my friends Helen and Darby. Far from the adventures of my childhood visits, which brimmed with sand fleas and the stink of organic ocean decay, you could call these trips reading retreats. We hole up in the condo during the heat of the day and read, then take our wild reading ways to the beach when the sun is low, hunkering under the wide umbrella, swiping Kindles and turning pages to the sound of the waves. Eventually the sun sets behind the condos to the west.

I don't mean to yuck anyone's yum, but every time I arrive at that stretch of road girded by high-rise condos with barely a hint of the glorious body of water on the other side, my heart clenches and I have to remind myself to breathe.

The last time I went to the Gulf I sat on the shore and watched the waves, the sand, the few shells that washed up at my feet, and apologized for my disappointment. "You are still the same, wild creature," I whispered to the vast body of water before me and the scrappy dunes behind. "It's not your fault, what they've done to you." I tried to release the heavy sorrow I feel for all the wild places into the evening breeze.

As I write, we're reeling in the aftermath of Hurricane Ian. The loss of life and livelihood on the Florida beaches is horrendous. I understand human suffering and I do not make light of anyone's pain.

But when I hear folks talk of the costs of building it back, I cringe. Build it back? Why not let wetlands be wetlands and do their job of protecting the shores? Why not go inland?

Maybe it's time to relent to the wildness of the Gulf, the oceans, the storms, the forces of nature that we will never control. Maybe it's time to reexamine exactly what it is we love about our beaches, and if that is what we've built there, maybe we never loved the beaches to begin with.

A Squirrely Gateway to Sanity
October 20, 2022

I'm mad for squirrels. I've contemplated dedicating a column to them previously, but squirrels? The pests in our attics, the thieves of our birdfeeders, and the tormentors of our dogs? Then I found out that October is Squirrel Appreciation Month. I may not be a traditional southerner in all respects but give me a theme and I'll run with it. Today we're here to give it up for our closest wild animal companion, the eastern gray squirrel (*Sciurus carolinensis*).

As a child I used to lie in the grass with my dog Abagail and watch squirrels in the trees. I knew those tail wags, those fits and starts, those chitters and yaps like the back of my hand. If only I had a pet squirrel, I'd think, staring at a nut-brown face, a pointed nose, a set of inky black eyes. I read about Squirrel Nutkin, Beatrix Potter's creation who lost a tail to an owl named Old Brown. Sometimes a tailless squirrel would visit the yard. "Squirrel Nutkin," I'd whisper.

Those longing memories give way to later ones of raising baby squirrels. It seems they were forever falling out of trees early. My brother Robert and I would take them home, craft elaborate nests of boxes, twigs, and cloth napkins, and feed them using eye droppers. Our rescue endeavors were carefully overseen by our parents, and many of our babies lived to adulthood.

When our squirrels were ready to be released, we had to wear blue jeans for the next several months or risk bloodied legs from adult squirrels scampering up our legs to our shoulders to say hello. Squirrels have excellent memory skills (it's a myth that they forget where they bury their nuts) and can long remember humans who've been kind to them.

As an adult I did not attempt to raise baby squirrels. I now appreciate my parents' lifestyle which afforded them the luxury of either bringing orphaned babies to their work sites or having someone home to tend to them. I never had that lifestyle, plus, it's illegal.

A few years ago, my husband Joe was out in the yard just before a big storm when a baby squirrel ran up to him and put his front paws on Joe's ankle. Joe immediately picked the baby up and brought him inside where we fashioned a warm snuggle box for him. Clearly weak and far too young to be out on his own, we contacted a certified wildlife rehabilitator. She explained that this behavior is not uncommon for fallen or orphaned baby squirrels. In some cases, the mother will return for her baby, but on this night, with a monster storm approaching, that was unlikely. Joe drove an hour away to deliver the baby to the expert.

Why will some people drive miles on a dark and stormy night to save a baby squirrel? From what well does compassion for helpless critters rise? It is my belief that if we foster our children's wide-eyed wonder of the natural world, and as adults remain cautious of the destructive human tendency to dominate and control everything that we come across, then the world will be a kinder, healthier place.

I am far from alone in my belief. One of my favorite environmental activists, Richard Louv, the author of the groundbreaking book *Last Child in the Woods* and coiner of the term "Nature Deficit Disorder," has written extensively about our relationship with the natural world and its effects on our emotional, physical, and mental health. In his latest book, *Our Wild Calling*, he explores the intersection of modern living with nature and offers glimpses of hope in mindful city planning, educational shifts, and acceptance of even the lowly coyote in our lives. I strongly recommend this book.

Back to squirrels. As adaptable and resilient as they are, squirrels did not venture into our urban areas on their own. In the mid to late 1800s, squirrels were imported into city parks to transform urban areas into oases of natural delights. Children were encouraged to feed squirrels, believing this would help them be less likely to engage in petty crime and vandalism. As city parks transformed into substantial greenspaces, so squirrel populations increased. According to Etienne Benson, author of the article, "The Urbanization of the Eastern Gray Squirrel in the US," "Squirrels were intentionally introduced in order to alter people's conceptions of nature and community… part of a much broader ideology that says that nature in the city is essential to maintaining people's health and sanity."

As we lose more and more of our wild spaces to unfettered development day after day, perhaps we should look to our little furry friend, the ubiquitous chattering squirrel, as our modern gateway to health, happiness, and sanity.

Beautiful Blue Corvids
October 27, 2022

I've always harbored a quiet love for blue jays, those oft-maligned backyard beauties. When folks speak harshly of them, I usually come back with a feeble, "But they're so pretty," to which no one can argue.

Well knock me over with a feather! I've just learned that jays are corvids, and I feel justified. Don't we all love corvids? Big-brained, intelligent, and adaptable, it's no wonder humans are enchanted with crows and their ilk, placing them in our creation stories and fables throughout history. I'm now proudly outing myself as a true-blue blue jay fan.

This fascination began in early childhood. Blue jays were the first birds I could identify by sight and sound. My brother and I found ourselves rescuing baby blue jays regularly. We would see jay birds divebombing our dog Bowzer out in the yard. Cued, we'd put on football helmets and rush out to wrest the fallen jays. Of all the little birds that came and went in our lives, most were baby jays.

The first bird I successfully raised by myself was a jay bird. I was in junior high, my parents were recently divorced, my brother was out perfecting the art of juvenile delinquency, and nobody else had time for a baby bird. I tended that little one with round-the-clock care until it was fully feathered and flying. At that point I named him Jay, having learned not to name a baby until it's past a certain vulnerability.

I watched awestruck as Jay developed the striking plumage of an adult. That mask, the intricate streaks of black and snowy white against such regal, brilliant blue (which isn't blue at all, but that's a lesson in physics)! What a magnificent animal he was! I assumed Jay was a male because at that time I did not know that male and female jay birds look the same, a rare occurrence in the bird world.

Jay soon developed a personality to match his good looks. He flew from room to room, stealing treasures and leaving bird poop. His vocalizations were jarring and loud.

"Time to put that damn bird outside," my dad said one day while attempting to read the paper while Jay tore the pages from his hands. I constructed a roost outside my bedroom window and allowed Jay to sneak inside despite his banishment.

To my delight he stuck around. He accompanied me on walks, bike rides, and even horseback rides. The magic of belonging to a wild bird aside, I could hardly believe his majestic beauty.

I left town for two weeks not long after Jay gained his independence. He never returned to his roost after that, nor did he fly down from the pine trees to say hello. I assume that he was taken in by the strongly social neighborhood jay birds. Although tame, he did not imprint on me; I did not expect a life-long bond. Still, his leaving was yet another brutal sadness.

Beautiful yes, but are they bullies? Smaller birds scatter when blue jays come to a birdfeeder, but the woodpeckers, squirrels, and sometimes cardinals run the blue jays away. In all these cases, smaller birds quickly return when the larger birds leave. It's true that blue jays are opportunistic omnivores, though most of their diet, about 75%, is vegetation. Although there has been talk of them killing smaller birds and nestlings it is now understood that this is a rare occurrence.

In fact, blue jays are beneficial. Their loud warning calls and defensive actions alert and protect other songbirds from predators. Their penchant for acorns helped the spread of oak and beech trees after the last ice age and continues to assist their coverage today. Blue jays can clear out a wasp's nest, queen and all, and they enjoy feasting on other pests as well.

Unfortunately, blue jay numbers are declining. Neonicotinoids, the same insecticides that are killing our bee and butterfly populations, are suspect. Banned in Europe, parts of Canada, and a few states in the US, the battle rages for the legality of this class of insecticide.

Autumn is the season for our raucous, boisterous blue jays. With the business of mating and child-rearing behind them, they are banding together to find food, chat about it loudly, and start building their winter caches. I'll be leaving peanuts out in our chemical-free wooded yard and hoping for new friends as blue jays, like squirrels, recognize the faces of those they trust.

A Donkey Tale
November 3, 2022

I can't believe it. Once again, National Mule Day (October 26) has come and gone, and I forgot to celebrate this noble beast. I suppose I can tweak things a bit and use this week's space to share a true story of a little donkey named Jeremiah who stole my heart many years ago.

Once upon a time there was a little Nubian donkey named Jeremiah who lived in a rolling pasture in Gold Hill, Alabama along with three horses and a herd of cows to keep him company. Jeremiah was what you call a "Jesus donkey," nicknamed for the story of the donkey who carried Jesus to Nazareth on Palm Sunday and who now carries the cross on its back for all eternity with a stripe down his back and across his shoulder.

Jeremiah did indeed have a cross on his back, but he didn't know it. All he knew was eating when he was hungry, drinking when he was thirsty, and sleeping when he was tired. It was a good life for a little donkey.

Jeremiah was long in the tooth as the old folks say, skirting around the word "old" as if it might come swooping back to rest on their own tired shoulders. Long past any sleek glossiness of his youth, his fur was thick and rumply. Occasionally the kind humans would brush it out to no avail. Jeremiah tolerated their attentions, but he preferred to be left alone to his own donkey musings.

One day, as he was walking toward the shallow creek for his morning drink of water, he caught a whiff of excitement in the air. His old ears perked up and the next thing he knew he was overcome with urgency. His little legs began trotting to the south end of the pasture. When he reached the barbed wire, he threw himself against it in a pheromone-driven frenzy, not even feeling the pricks and jabs in his thin skin.

Finally, the barbed wire relented. He was free! Jeremiah didn't actually care about freedom. He was, in fact, still captive but now his captor was a scent, and that scent was all he cared about in this live-long world.

It turns out that the smell that lured old Jeremiah away from his home pasture was that of a young mustang filly in heat. Jeremiah, being a Jack donkey (that is, fully intact), had no choice in the matter but to find her, and after following the road for a mile or so, he did.

There was much ado when Jeremiah reached the young mustang filly. Following a command larger than a mountain and older than time itself, he began to prance about wildly, finally throwing his front hooves up onto the back of the little horse.

Just around the corner, a human caught sight of the goings on. She raced inside the house to a phone. The next thing he knew, Jeremiah was being literally manhandled away from his prize and into the back of an old Chevy pickup truck. As the truck rambled down the highway, the scent and the memory of the little mustang filly faded.

Exhausted and a bit banged up, Jeremiah was happy to be returned to his own pasture where the three horses trotted up to meet him, nostrils wide with curiosity.

In the morning Jeremiah returned to his creek. He'd worked up a powerful thirst by now. The water was cool and clear and felt good on his tired legs, which were bruised and sore from his adventure.

Thus ends the tale of Jeremiah the little Nubian donkey, who was found quite dead later that morning, his feet still in the creek, his head resting peacefully on a mossy clump of dirt.

And thus begins the story of little Alice the Mule, who made her debut into this world about 11 months after Jeremiah's fateful adventure. Fuzzy-eared, long-legged, wide-eyed and bright as a beetle, one look at that baby mule could soften the heart of the hardest cynic.

I had the honor of meeting little Alice the Mule within her first week of life. She sniffed my hand and nibbled my shirt, and I rubbed the soft red hair on her neck.

"Hello little Alice the Mule," I said. "Let me tell you about your daddy. His name was Jeremiah."

Happy belated National Mule Day. May we all learn from the patience, courage, strength, and intelligence of these delightful animals.

Rattus Rattus and Friends
November 10, 2022

What happened to fall? It sure is hot outside. In that quick cool-weather tease, we celebrated the arrival of a highly anticipated annual event at our house: turning on the heat! Alas, only cool air blew from the vents. A quick $100 visit from a repairman determined that rats had built a nest directly over the heater's pilot light.

"Empty nest, right?" I asked, concerned for the baby rats. We'd discovered evidence of their presence this summer and gone to great lengths to trap and release the unwelcome rodents far away where they'd likely be eaten by a bird of prey, as the circle of life intended.

Despite the fact that rats cause roughly $20 billion to US homes, businesses, and agriculture annually, I am quite fond of them. I am not alone. Their keen intelligence, cleanliness, and bright eyed, twitchy-nosed cuteness have made rats increasingly popular as pets. They are making headway in other areas as well. From sniffing out landmines to providing insights on empathy, rats are highly valued for their fine-tuned skills and senses.

There are 56 species of rats. The ones that invade our homes and build inconvenient nests are the Norway rat (*Rattus norvegicus*), commonly known as the brown rat, and the roof rat (*Rattus rattus*), or black rat. Roof rats, as their name implies, enjoy nesting in roofs and attics while Norway rats nest closer to the ground.

Pet rats are called Fancy rats, so named because folks fancy having them as pets. Despite the fancy name, pet rats are simply a domesticated variation of brown rats. When we were little, my brother Robert and I had pet rats. Robert named his first rat Willard and I named mine Baby, which tells you all you need to know about our distinct personalities. Baby and Willard were tooth-grindingly cute. They had delicate ears, shiny black eyes, and the sweetest pink noses and paws. Curious and calm, they were as interested in us as we were in them.

My good friend Dana decided that she too needed a rat. In no time, Baby had a new friend: Ratmaroff. Dana and I carried Ratmaroff and Baby around like toys. They joined us in our never-ending games of Monopoly, sat patiently on our shoulders while we played Jacks, and accompanied us on weekly walks to Zippy Mart where we spent our allowances on candy. Rats, like dogs, are social critters and require a great deal of interaction with their humans to keep them happy. We were excellent rat owners.

As parents, Joe and I decided our girls needed a pet rat, and Susan the Rat came into our lives. Thank goodness Aunt Susan the Human knows that having any pet as a namesake in our family is a great honor.

Susan the Rat had a cage, but she was free to roam the house and roam she did. Every night she'd join us on the upstairs couch for book-reading, then appear on the downstairs couch for movie-watching without using the actual stairs. We never knew the exact location of her secret tunnels. She'd visit her cage now and then for food and water but spent most of her time hanging out with us.

As lovely as they are as pets, I concede that wild rats are a nuisance, and it's probably best to discourage them from nesting in our homes. Let's talk about rat control. Glue traps are off the table, period. They are excessively cruel, barbaric, and inhumane. For the love of all that is holy and right, please do not use glue traps for any reason.

Rat poison is also inhumane, causing a slow death as organs shut down. The consequences of rat poison reach far beyond the poor rat, however. As poisoned rats grow weak and dehydrated, they often seek water out in the open, becoming easy victims for animals who prey on them and become poisoned themselves. Rat poison wreaks havoc throughout the ecosystem, particularly on birds of prey.

Be aware that pets and children can also become victims of rat poison, and depending on what kind of books you read, errant husbands as well. Rat poison is banned in California, with other states expected to follow suit.

Snap traps are effective and relatively humane for those who can stomach killing rats. We use kind traps, as we cannot stomach killing rats. The risk with kind traps is bonding with your critter while toting it to that faraway place.

The best rat control is right under our noses: rat snakes. Humane, safe, and beneficial to their ecosystems, rat snakes are exceptionally efficient at rodent control. They can sniff out vermin under piles of debris and in burrows, basements, and other hidey-holes. The good news is that if you have rats, you have rat snakes! All you have to do is let them live.

With the empty rat nest removed, the pilot light lit, and the end of Daylight Savings Time here, I'm ready for winter. Bring it on, Alabama. This heat grows tiresome.

Ancient and Imperiled Gopher Tortoises
November 17, 2022

Big news! The Gopher Tortoise Council held its 44th annual meeting last weekend in Freeport, FL. Gopher tortoise champions and researchers gathered to discuss recent regulations, status updates, and best conservation practices for these important reptiles.

I have a special place in my heart for the Gopher Tortoise Council. Those three words alone transport me back to childhood. One of my first exposures to activism was tagging along with my dad to the Gopher Tortoise Council meetings held at Auburn University back in the 1970s.

In adulthood, other than wishing I still had my original Gopher Tortoise Council t-shirts, I didn't give much thought to those tortoises until I read Janisse Ray's *Ecology of a Cracker Childhood* (Milkweed Editions, 1999), a book which opened my eyes to the complex, gorgeously choreographed longleaf pine forest ecosystem. I began to appreciate the gopher tortoise anew, as well as the red-cockaded woodpecker, the eastern indigo snake, the bobwhite quail, and the many other species living in the longleaf pine forest.

The gopher tortoise (*Gopherus polyphemus*) is one of the oldest living species on the planet, and the only tortoise native to North America this side of the Mississippi River. They are notable for their large paddle-like front limbs which they use to dig elaborate burrows, a behavior acknowledged in their very name: "gopherus" after the burrowing gophers, and "polyphemus" after the mythological Greek cave-dwelling giant. Their burrows, which are typically around 15 feet long but can be as large as 40 feet long and 10 feet deep, are home to over 360 species. This earns the gopher tortoise the distinction of being a keystone species, that is, a species whose demise would significantly alter an ecosystem or cause its collapse entirely.

The gopher tortoise range includes the southeastern coastal plain, well south of our home in Lee County, Alabama. There was a day, however, when a gopher tortoise showed up in our neighborhood.

Our daughter Anna, age four at the time, found a funny little turtle in our wooded yard. She named it Speedy, and after playing with it a while, turned it loose to carry on its natural business. We assumed she'd found a box turtle and gave it little thought.

The next day when we went to fetch her from her pre-school, located in a house about a mile from ours, she excitedly introduced us to Speedy. Speedy was no box turtle. He was a young gopher tortoise who had miraculously found our Anna not once but twice in two days!

Despite the fact that my dad wanted to keep the tortoise as a pet (rules did not apply to him), we located its rightful caretaker, a young researcher with Auburn University. He was greatly relieved to have his valuable specimen returned safely.

"How is it that this little fella found Anna twice?" I wondered.

"Gopher tortoises are like that," my stepmother Janie answered, as if everyone knows their mystical powers.

This gopher tortoise tale had a happy ending. Unfortunately, the current state of gopher tortoises is concerning. Although they have survived for 60 million years, we humans have managed to cause an 80% decline in their numbers over the last 100 years through the destruction of the longleaf ecosystem, land fragmentation, fire suppression, roads, urban sprawl, development, and other short-sighted activities. Still, in October 2022, the US Fish and Wildlife Service determined that some populations do not merit protection under the Endangered Species Act. This decision remains contentious.

Isn't it astonishing that we humans can be so destructive to the very planet that sustains us? It's as if we somehow think we have the right to destroy the livelihood of other species without a backwards glance. It's as if we've lost sight of the fact that we are part of the living world, separated from our environment, including other creatures, by mere cell membranes.

I had the pleasure of hearing the aforementioned environmental activist and writer Janisse Ray at the Arts at the Confluence 2022 celebration at First Church in Atlanta recently. She closed her talk with this quote from poet Adrienne Rich, bringing herself and her audience to tears:

"My heart is moved by all I cannot save: so much has been destroyed. I have to cast my lot with those who, age after age, perversely, with no extraordinary power, reconstitute the world."

Amen, and ever gracious thanks to that lot, including those who look after the well being of animals like the Gopher Tortoise.

The Greatest Animal of All Time
November 24, 2022

We're a dog family. Right now, I'm sharing a couch with six pooches. Every time my husband Joe walks by six heads rise, 12 ears prick, 12 eyes sparkle. Then all gazes turn to me. "Will she get up?" they seem to ask before Joe disappears into the next room. Satisfied that our couch lounging will continue, they lower their heads, relax their ears, and close their eyes. I am distracted by their charm. Dogs are, hands down, my favorite animal on this planet and beyond.

It's almost Thanksgiving as I write, and when this column goes to print it will be Thanksgiving Day. While research supports the hypothesis that gratitude is not only nice but beneficial to our mental and physical health, I don't need a day set aside to give thanks for dogs. Every day of my life I am overwhelmed with gratitude for the spark that kindled this beautiful friendship some 30,000 years ago.

That spark was likely lit by members of a species of wolves, now extinct, who overcame their fear of hunter-gatherers to scavenge bones and other tasty left-behind morsels. What happened next? The story that wolf-pups might've been captured and raised by these early ancestors has been debunked. Among other things, anthropologists assure us that life was too difficult for people to opt for bringing in yet another mouth to feed.

It's far more likely that women were responsible for building what has become a sacred trust. As strong, hungry wolves began following men out on the hunt, the younger and/or weaker wolves stayed behind with the women. A gentle look here, a kind word there, and bonds were formed. The wolves who had stronger connections with women were more likely to be accepted in daily life.

Over time both humans and wolves changed, and here we are, sitting on the couch together tens of thousands of years later.

The evolution of humans and dogs is considered "coevolution," a situation in which two distinct species develop adaptations that benefit one another. The relationship between pollinators and flowers is the classic example, characterized as "commensalism," or mutually beneficial.

What exactly are these friendly beasts beside me now? Are they little wolves? Although the same species, *Canis lupus*, dogs are not simply tame wolves. *Canis lupus familiaris*, domesticated dogs, are one of over 30 subspecies of the gray wolf. We often say, "Dogs are descended from wolves." It is helpful to think of this path as a divergent branch, a branch, rather than a hierarchal, straight line.

Unlike wolves, dogs are capable of initiating and holding eye contact with humans. This may seem trivial, gazing lovingly into the eyes of one another, but in fact it produces the hormone oxytocin, nicknamed "the love hormone," in both humans and dogs. In a beautiful open-feedback loop, dogs also produce more oxytocin in response to the smell of human oxytocin, and the gazing cycle continues. Be careful not to fall down the rabbit hole of research on "The Dog-Human Gaze." You might miss Black Friday.

Dogs have developed facial muscles not found in wolves which enable them to make "puppy dog eyes." This precious expression benefits the survival of the species, with humans responding favorably to cute faces. Additionally, dogs have rounder eye sockets, shorter snouts, and floppier ears than wolves. Dogs are not just tame; they are domesticated, which means they have undergone genetic changes which distinguish them from their ancestors.

Sometimes in moments of utter contentment, like right now with these six dogs, it feels like the history of life on this planet has conspired to lead to this exact point. Hallelujah! In fact, we're here due to a bunch of random mutations coupled with natural and artificial selection for favored traits, and although this moment seems like the pinnacle, all those actions are still at work. Both humans and dogs are still coevolving, our behavioral and physical adaptations leading the changes.

I could write a world of words about all the ways dogs are being trained to detect this and that to the benefit of humanity, and the research behind companion dogs for personal and institutional benefits, but my dogs would tire of this. They are all gazing at me now, and I must stop writing and return the gaze. I have no choice. It's biology.

Before I go, I must give a shout-out to the dogs with whom I have shared a home: Nick Dog, Abagail, Bowzer, Ziggy, Boo, Ugly, Gruff Chester, Ricky, Waldo Pepper, Otis, Pee Wee, Garth, Kennerly Punk, Lucy, Phoebe, Houndy, Poodle, Fidel, and Che; to our current pack: Lola, Peedie, and Bernie; to our grandpups Flea, Pru, Fig, and Tulip; to Susie and Ice, the cousin dogs; and to all the dogs everywhere. Thank you for making the world infinitely better.

And to the humans, Happy Thanksgiving to you all!

On Nostalgia
December 8, 2022

Last night at a gathering of friends I conducted some festive research and asked people their thoughts on nostalgia. To a person, they paused, eyes focused on something unseen by the rest of us, then shared a wistful memory. Intentions aside, we had all brought along Nostalgia as our Plus One to this holiday party.

I am often overwhelmed by nostalgia, with Christmas as a powerful trigger. The ghosts of my mother and my little-self drag me out of the present with each step of the decorating. We are again tiptoeing into the Lutheran church with candles in our hands to sing carols at midnight, gathering magnolia and holly boughs to decorate the mantle, polishing silver bowls and tying red ribbons on candlesticks, my mother and me together, just us.

In the past few years this intense nostalgic crush started happening in the summertime, particularly when I cut fruit. One June day two years ago I could barely slice an apple. As soon as the blade touched the peel, I was transported to summers of my early childhood, running errands with my mother, looking for the crispest apples and sweetest plums, stopping in Crest Drugstore for jet-black India ink for her fountain pen. I called her to tell her what my brain was doing.

"You're having a Proustian experience," my mother said. She was such a fan of Marcel Proust that she named her Rollator Walker Marcel. I thought about this as I bit into a perfect plum.

My mother died four months later. Decorating the tree that year was soul-wrenching and cathartic. There we were, back at the flower store buying green wire and Styrofoam cones. Together. Honoring our ritual of hanging 2000 strands of icicle tinsel on the tree one by one gave me 2000 discrete moments to be alone with my mother. Was I sad? Was I happy?

What is nostalgia?

The word comes from the Greek root "nostos" meaning "return to home," and "algos" meaning "pain." Coined by a Swedish doctor in 1688, it described the homesick condition of Swiss mercenary soldiers. Doctors concocted a cure of leeches, opium, and a return to the Alps to remedy this mysterious ailment.

Rather than taking this esoteric cure, we welcome nostalgia today, especially during the holidays. We beckon it with scented candles and Spotify playlists while decorating the house; we revel in it as we cream the sugar and eggs for the pie we make only once a year.

As with most things worth pondering, nostalgia is complicated. Brain imaging shows that four distinct areas light up when we experience it's pull. Paradoxically, the brain perceives nostalgia as both novel and familiar, giving it a double-edged superpower on our emotions. There must be some evolutionary advantage for this mystical longing, but what?

Maybe it's like glue, keeping our present selves psychically connected to our past selves in a way memory alone cannot. After all, a body is merely a collection of atoms squashed together at a given moment in time, with atoms constantly flowing in and out. Despite DNA's persistent instruction, we are in a constant state of flux. Something has to hold us together.

Nostalgia also connects us to things we've never experienced. Did I shop on 5th Avenue, twirling in the snow, elbows bent with the weight of my Saks 5th Avenue and FAO Schwartz bags? Did I ride on a sleigh behind a team of Clydesdales, bells chiming in my ears, through a wintry landscape? Did I play a rousing game of football with my cousins after a holiday feast? No. Yet nostalgia leaves me all atingle with a fuzzy longing to return to these events from an imaginary past.

Whatever it is, nostalgia binds us to something. A bond to other people and times is indeed advantageous to our health and survival as individuals and as a collective. Perhaps this is nostalgia's advantage.

It's December again, and here we are. The ornaments are on the Christmas tree, and it's time for the 2000 strands of tinsel. I look forward to walking hand in hand with my mother again during this days-long ritual.

I'm not sure I understand the evolutionary advantage of nostalgia yet, but I am sure I understand the balm it has on the soul.

A Place in the Heart for Protists
December 15, 2022

It wouldn't be Christmas without sparkly pictures of amoebas, paramecia, and *Euglenas*, would it? Perhaps I should explain.

I have a soft spot in my heart for some of our tiniest living things. I call them the Big Three of the Protist Kingdom: the amoeba, the paramecium, and the *Euglena*. I can't recall the very first time I first heard of these protozoa, but my love for them began in 7th grade.

My life science teacher was small, soft-spoken, and kind. I think her name was Mrs. Smith. As well as being a patient teacher, she gave me a ride every Friday to the farmhouse in Gold Hill where I spent weekends on my horse, thus tying her inextricably to the place I loved the most in my formative years.

One day we were dissecting frogs in science class. My frog had a bellyful of undigested crawdads. Like the girls beside me, I squealed in disgust when I sliced the stomach open and saw the leggy arthropods. Mrs. Smith materialized over my shoulder in an instant. "Don't act like that," she said quietly, just to me. "That's not who you are."

It's well known that 12 is a precarious age, one in which the self is paradoxically trying to break free and to belong to something at the same time. To an awkward, sensitive child of recently divorced parents, her words gave me the clarity I needed to put down my stake in this shaky world. In that moment I became a scientist, with biology my new true love.

One day Mrs. Smith introduced us to three microorganisms, the aforementioned Big Three. We drew pictures of them and examined them under microscopes. I'd known ponds and lakes were full of microscopic beings, but I was unprepared for their charm.

"They're protists," she said, and I wrote the word in my notebook, ready for more.

For those of you who have forgotten, here is a quick primer. Members of the protist kingdom include the amoeba, the paramecium, and the *Euglena*, as well as algae, marine diatoms, and slime molds (swoon). All protists live in moist or watery environments. Some protists consume food, some photosynthesize, and some do both. Protists vary widely. In fact, Kingdom Protista is known as the Grab Bag Kingdom. If an organism isn't an animal, plant, fungus, bacteria, or archaebacteria, it gets tossed in with the protist lot.

I colored the blob on my paper and labeled it "Amoeba." It had oozy extensions, pseudopods, which it used to move and to engulf food. The paramecium was more complex, with a flower-shaped contractile vacuole, a macro- and a micronucleus, and a fold called an oral groove. It was covered in cilia, tiny bristles which it used to move through the water. My book said the paramecium was "slipper shaped" and it used its cilia like "oars." My head reeled in visions of paramecia boating through lush ponds.

But oh, behold the *Euglena*, the only one with a capitalized and italicized name. *Euglenas* come ready for Christmas with their green bodies and big red, eyespots which detect light and guide the whiplike structure, the flagellum, to propel the *Euglenas* toward light in order to photosynthesize. Like a tiny person holding the reins, I thought when I saw how the flagellum branches off at the eyespot.

All this was going on in ponds across the world. Amoebas, *Euglenas*, and paramecia swimming about, moving with their specialized locomotion structures, each storing food, ejecting waste, reproducing, all within a single, solitary cell. My brain tingled to think on it.

I have never quite recovered from my shock and awe of this tiny world. When I taught life science, I imparted my enthusiasm to my students. We drew large, simple sketches of protists and decorated them for Christmas. A bit of curling ribbon makes a fine flagellum; gold glitter enhances an otherwise dull amoeba. Sometimes we combined this activity with writing protist short stories and hosting protist beauty contests.

Much has happened since that day in 7th grade when the tiny world of microorganisms was introduced to me. Industry has found uses for *Euglenas* ranging from biofuel to nutritious smoothies, and the clickbait of "Brain Eating Amoebas!" has crossed our social media feeds more than once. Textbooks still use the word "slipper-shaped" to describe paramecia, which have become the darlings of the classroom due to their ease of care and lack of pathogenic qualities.

One thing that hasn't changed is the nature of middle-school kids, who still clamor and clang loudly to be seen for who they are while hiding beneath layers of icy cool. Here's to the Mrs. Smiths of the world who help them find their way.

Seasons, Solstices, and Old Man Winter
December 22, 2022

Happy second day of winter! Yesterday was the winter solstice, the day when our home planet's northern hemisphere is tilted as far away from the sun as it gets on its journey round our personal star. Starting now, we get a little more sunlight in our dreary daytime hours. It makes no sense, but during winter, our days grow longer.

The definition of seasons is peculiar. It always seemed to me that the solstices and equinoxes should mark the middles of the seasons, not the beginnings.

Well, not always. There was a day, or a season of my life called childhood, when I didn't think about these things a'tall. Summer was marked by the months with no school. Fall was back to school, winter was cold, and spring had flowers.

Now I confess, I didn't actually understand seasons until I took a world geography class with Dr. Dorman at Auburn University. Dr. Dorman commanded full attention. He stormed about the classroom with a cigarette in his mouth, never tapping the ash. He tapped his heels, though; Dr. Dorman wore taps on his boots which made quite a racket when he slammed those boots on the seats of empty chairs. He also carried a stick. Once a brave or possibly dense kid opened up a newspaper in the middle of class. In a flash, Dr. Dorman snapped his stick down and whipped that newspaper into the air where he grabbed it, quick as a wink, with his free hand and continued his lecture.

"When the northern hemisphere is tilted furthest away from the sun," Dr. Dorman boomed, "the sun's light never reaches beyond the Arctic Circle!" He slapped the big map on the wall right at that parallel. The map shook slightly in response.

Maybe it was Dr. Dorman's urgency. Maybe it was a developmental milestone. For whatever reason, it was then that I actually understood the position of Earth in space and what all the named lines on the globe meant. The Tropic of Cancer and all the others were just marks where the sun's rays shone on the curvy planet: direct, indirect, not at all.

Suddenly, I loved geography. I papered the walls of my college bedroom with maps. I shone a desk lamp on my globe and pondered the Tropic of Capricorn in new light. I marveled that at those four places in our revolution around the sun, thanks to our 23.5-degree tilt, we have four days a year of extremes. No wonder the ancient civilizations got excited on the solstices and equinoxes! And this in the days before the Dr. Dormans of the world had the benefit of science. Isn't it all just spectacular?

But back to the definition of seasons. If winter begins on the "darkest day" and all days following get longer until we reach the "longest day," marking the beginning of summer, why does everyone say that in winter the days grow short?

It turns out we have another definition of seasons.

Meteorological seasons are based on the months during which temperatures are typically most extreme. Meteorological winter began on December 1st and lasts through February. I feel justified.

Whenever it started, Old Man Winter is here. The sky is dreary and gray, the limbs of the trees are bare, and the red cardinals show up bright as berries against the empty sky.

This year, rather than frantically wrapping the presents for the grandchildren and loading the car with Christmas fancy, we're ignoring all things Christmas and heading out to Edisto Island, SC, where we will hole up in a house overlooking a salt marsh and watch birds and dolphins for a week.

If ever there is a place where you can feel your connection to the planet, it's on an island. Staring at a sunset you can see Earth turning, the sun disappearing to your view. Staring at a salt marsh in the morning, you can see the golden blaze approaching across the watery grass.

Solstices, equinoxes, Christmases, birthdays, any old Monday, it's all just a blink and a twitch to the planet. Earth doesn't give a tinker's rip how we define "seasons."

Merry Christmas and Happy Holidays to all! May the light of the sun warm your cheeks as we head toward our next planetary extreme, the spring equinox!

Dolphin Walks on Edisto Island
December 29, 2022

I don't think there's a single solitary person on this big blue planet who doesn't love dolphins, myself included. Like all healthy minded kids, at one point in my childhood I wanted to be a dolphin trainer. Questionable ethics and cruelty aside, that goal was but a blip as my preferred dolphin encounters turned to spotting them in the wild.

My first memorable encounter with dolphins in their natural element was when I was 14 years old, deep-sea fishing in the Gulf of Mexico with my friend Saralyn. These days I can't bait a hook, let alone bring a fish out of the water, but I was younger then and believed the old salts when they said things about it not being cruel to the fish.

Saralyn and I were tending our giant rods and reels when a pod of dolphins joined our wake. Delighted with their playful leaps and cunning smiles, I decided deep sea fishing was the life for me! That is, until I felt a queasiness that would not subside. A gnarled and gruff fisherman took me down into the bowels of the boat.

"Sit," he growled.

He pushed a greasy cheeseburger, onions and all, in front of me and popped the cap off a Budweiser. "Eat. Drink," he grumbled, handing me the bottle. I, still believing old salts, did as told. I spent the rest of the ride in the fetal position, quite green around the gills. Still, I'd seen dolphins up close in the wild and from then ever after harbored a fondness for the idea of deep-sea fishing, though I never tried it again.

After that, my encounters with wild dolphins were chancy and scarce until we discovered Edisto Island, SC, 18 years ago. Edisto Island is part of the ACE Basin, one of the largest wetland ecosystems on the Atlantic, formed by the confluence of the Ashepoo, Combahee, and Edisto Rivers. A good 350,000 acres of preserved marshes, hardwood forests, and river systems makes it a rich haven for critters.

We didn't know any of this when we stumbled upon this mysterious spot, located at the end of a 25-mile highway canopied with live oaks and Spanish moss. On the last day of our first visit, someone asked if we'd been down to "the point" to see the dolphins.

"The point" is the place where Big Bay River snakes through salt marshes, joins the South Edisto River, and flows into the St. Helena Sound. There the water is calm, the tidal flat is blanketed in shells, the pluff mud oozes stinky and fecund under the crunchy marsh grasses, and the sunsets blaze outrageously. But the crowning jewel of this bit of heaven is the presence of a permanent pod of Atlantic bottlenose dolphins (*Tursiops truncates*). We frequently stroll alongside them as they head out to or return from the open sea, feeling blessed beyond measure to enjoy our dolphin walks.

One day I took a kayak tour with a naturalist named Meg. We paddled into the rivulets of the salt marsh, marveling over the solid mass of fiddler crabs on the muddy banks and the hungry herons attempting to eat them. Suddenly two dolphins came speeding right toward me. I feared they had gone mad. Then they hurled themselves onto the riverbank, writhing and twisting furiously, and just as quickly, slid back down the mud into the water and swam away.

"You just witnessed strand feeding," Meg said. "You're lucky."

She went on to explain that strand feeding is a rare feeding habit in which dolphins work together to herd fish onto the banks of a river or mudflat. They then momentarily "strand" themselves while snapping up their prey. Their mouths full of food, they writhe and twist about, forcing their beached bodies to slide back into the water.

How had I gone my whole life not knowing about this, I wondered. It turns out that the only dolphins known to engage in strand feeding are those living on the banks of Georgia and South Carolina. Incredibly, it is not innate, but a behavior taught to each generation.

This year my husband Joe and I, plus our three dogs, spent Christmas on Edisto Island. We braved the frigid 30 mph winds for a few dolphin walks at the point. There they were, leaping out of the water, showing off pale pink bellies and wide smiles, just a few yards in front of us. Lola the Wonderdog barked in glee then jumped into the water and splashed with them.

Even during a record-breaking cold week, the wild and untamed nature of Edisto Island is a balm to my soul. I look forward to our next visit and the opportunity for dolphin walks along the fertile shores of the St. Helena Sound.

Happy New Year to all, and may your 2023 be filled with wonder.

Bull Batkin!
January 5, 2023

While we spent a lot of time outside growing up, family neighborhood walks were a rare occurrence. That's probably why I remember them with the same tingling sensation usually reserved for holidays and birthdays. On these infrequent strolls, I kept my place between my parents trying to match their long strides with my short legs, while my brother Robert ran in circles jumping into kudzu patches, climbing trees, and peeking into neighbors' windows. Abagail, our perfect mutt dog, trotted ahead of us, wagging her tail in time with her step. Daddy would throw little stones onto the street so hard they'd spark, which was almost as magical as the muffled adult conversation streaming over my head.

My favorite part of the walks was when the bats came out. Illuminated by flickering streetlamps, they'd circle tightly then fan out on their evening hunts. Daddy would point at them and yell, "Buuuuuuull BATKIN!" then grab my hand and run. Physics tells me it's impossible, but in my memory, I was airborne as we flew through summer nights with the bats.

Perhaps this is why I was so moved to find an abundance of bats out in full force in the fields of our local dog park a few months ago. After a chance viewing, Joe and I timed our subsequent evening dog walks to catch the spectacular ballet. Huge grasshoppers leaped out of the tall grass, a meaty banquet for nearby insectivores. Multitudes of bats and nighthawks appeared and swooped overhead, backlit by a golden setting sun.

As the bats and nighthawks dove and zig-zagged in the darkening sky, I heard my dad's silly call, "Bull BATKIN!" as clearly as if he were standing beside me. Isn't it amazing how our brains can hold onto the voices of our loved ones, playing them back at the most unexpected times? I'm the only one left of my original family of four, and I savor these memories when their voices, like bats, capture my attention with their beautiful outflights.

I wish for all children, all humans, to have magical bat-filled nights, but sadly, over half of the 45 species of bats in the US are endangered. It is estimated that over six million have died from white-nose syndrome since 2006. Habitat loss, climate change, and wind turbines also threaten our bats. Researchers have found that spinning the wind turbines at higher speeds can significantly reduce bat deaths. Unfortunately, the other obstacles are more difficult to combat.

Does it matter? Responsible stewardship of the planet aside, most folk are familiar with the benefits of bats as pollinators, seed distributors, and insectivores. The Journal of the Association of Environmental and Resource Economists recently published research concluding that the loss of bats is costing American farmers $495 million a year.

It's not just bats. The devastating reality of species loss, with some estimating that it is occurring at 10,000 times the natural rate, is overwhelming. Real solutions to this terrifying decline in biodiversity lie in global policy changes and a completely new vision of consumerism. It's enough to make you want to curl up in the fetal position. But that won't help.

We can't change the world in one fell swoop, but we can make a difference to our backyard bats. We can start with welcoming them with bat houses, including dead trees, and clean water sources. Reducing excess lighting, minimizing tree and brush clearing, planting native plants to encourage pollinators, and avoiding pesticides not only help our bats but all our native backyard critters. Should we find unwanted bats in our belfries or homes, we can seek out safe and humane removal services (a bat in the house is no cause for alarm, but you might need assistance with a colony).

Out in Texas, the Houston Humane Society and Texas Wildlife Rehabilitation Coalition recently went a bit further, rescuing over 1,500

Mexican free-tailed bats who fell from their roosts during our historic Christmas freeze. The fact that that the same species that partakes in dogfighting can also tend to hypothermic bats sends my brain into a tangle. There are so many soulful people in the world, including the good folks here who run the Alabama Bat Monitoring and Conservation Program headed by Nicholas Sharp (Outdoor Alabama).

Thank you to those who rescued the bats in Texas, to Nick Sharp and the others working to protect our local bats, and to all of you who do your part in large or small ways to share space with wild animals.

Oh, one more thing. "Bullbat" is actually a nickname of the common nighthawk. I'm certain my dad knew this, even though he let me go my whole life thinking a "bullbat" was one or our local bat species.

My Sweet Tooth, My Microbiome, and Me
January 12, 2023

When we were little our parents told my brother and me that there was a battle between the germs and the leukocytes going on in our bodies. "Send down the sugar!" the germs cried daily, while the leukocytes called in response, "Send down the vegetables!"

I listened, and in adulthood I can claim pretty good health, yet I have a bit of a sweet tooth. Okay, a beast of a sweet tooth. In fact, my personal research supports the hypothesis that the only way I can stop eating Reese's Miniatures is to eat them all.

Why, I ask myself over and over, can I not resist these sweets? Surely it is some moral failing on my part. What self-respecting adult can devour Valentine hearts, candy corn, and jellybeans?

Imagine my delight when I stumbled upon news today that I might be able to blame something other than my will power. This childish craving for sugar could be the fault of my gut microbiome.

Cocktail party conversation has taught us that we house more microscopic beings, or microbiota, in our bodies than we do human cells. Research from the 1970s put that ratio at 10:1, but newer findings give only a slight edge to microbes, which is still astounding. Most of these little friends live in our digestive system and on our skin. Altogether, this wellhouse of bacteria, fungi, protists, archaea, and viruses weighs about 3 pounds in a typical adult. (If this is not your usual cocktail party conversation fodder, you might be going to the wrong cocktail parties. Or maybe I'm going to the wrong cocktail parties.)

What on Earth is this microbiome up to? Of course, digestion is one of its primary functions. We've all experienced the unpleasant tummy troubles when taking antibiotics, due to the good bacteria in our gut being killed off along with the problematic bacteria.

But oh, there's more. There's so much more. Exploring new insights into our microbiome is almost as much fun as a walk in the woods with a toddler. Behold the wonder and the awe of the tiny world of our inner ecosystem!

Here is a quick run-down of the latest findings. A healthy microbiome can help control type 2 diabetes, relieve symptoms of IBS, improve bone strength, decrease anxiety and depression, help the body maintain a healthy weight, relieve stress, possibly prevent Parkinson's disease, improve sociability, and decrease behavioral disorders in infants. This is only a partial list.

And that sweet tooth I mentioned? New research out of California Institute of Technology finds that specific gut bacteria may suppress "hedonistic" eating in mice. When researchers knocked out the microbiota of mice using antibiotics, they binged on sugar. When they replaced the microbiota via fecal transplants, the mice returned to their healthy eating behavior.

Well knock me over with a chocolate chip cookie. It's the lack of some tiny organism inside that makes me grab the 20th Hershey's kiss. Now I need to know what bacteria I lack in my gut and how I can get it back, preferably with a simple, easy-to-swallow pill and not a fecal transplant. While the Caltech team has identified the bacteria, as it relates to me and my sweet tooth, "further research is needed," blah blah blah. In other words, there's still no magic bullet. Yet.

What am I to do? I can take better care of the microbiome I already have. The recipe hearkens back to what my parents told me, "Send down the broccoli!" vs "Send down the cookies!" A diet rich in colorful, plant-based food; lean meats; healthy nuts and oils; and one low in sugar and processed foods promotes a healthy microbiome. Locally produced and seasonal foods are healthier than foods from far away that have sat in cold storage and traveled great distances to reach us. Added sugar, alas, is a major disrupter in the health of our microbiomes.

Our behaviors can also promote healthy tiny worlds within. Socializing with those we love, which floods our brains with oxytocin and serotonin, is also good for our microbiota. Such community activity invites us to gather bacteria from friends, a bonus for the gut, as does being outside, exposing ourselves to bacteria from Mother Nature.

There we have it, the key to good health and to taming the sweet tooth: eat real food, play outside, and spend time with friends. It sounds a lot like what my parents taught me.

Mule Team Memories
January 19, 2024

I didn't get home until dark last Thursday, the day the strong storms came through and winter tornadoes made their mark on Autauga County and the city of Selma. Relieved for the pass, I huddled with Joe and the dogs on the couch and watched news clips of the devastation for those in its wake, horrified at the brutal and indiscriminate destruction of the day's storms.

When I took the dogs out to relieve themselves before bed, my nose was greeted by the delicious acrid scent of freshly cut pine. A tree's down, I thought, looking up into the starry sky, now clear and calm, taking in one of my favorite smells. No tree-shaped air-freshener, no Mrs. Meyer's pine-scented soap, no Glade "Pine Wonderland" candle can come close to the real deal.

It turns out a neighbor lost two large loblolly pines. They now rest comfortably on the ground in our yard, where grubs, worms, snakes, rabbits, fungi, lizards, insects, and other forms of life will take up residence. Grateful to have escaped roof damage, I walked around the fallen branches and was quickly transported to childhood, back to a time when a tornado came through Auburn and changed the face of our town completely.

I wish I could tell you the year, but my childhood memory knows no linear paths. We attended school at Lee Academy, now Lee-Scott Academy, which at the time was housed in what I would describe as temporary shelter, a few rows of metal buildings on a hill. This little cluster of classrooms was nothing like the tornado-safe buildings of today's schools.

On this day in my memory, our mom came to pick us up ahead of a storm. As we drove home my brother Robert and I stared out through the back windshield of our dark green Chevy Impala, watching trees fall against the darkening sky.

At home, Mama rushed us into the house where we sat against an interior wall. We watched as tall pines outside swayed, snapped, and fell to the ground with house-shaking thuds. When it was over we couldn't wait to go out and see the wreckage. Our parents put the fear of God into us with tales about "live wires" that could snap around like angry snakes and kill us instantly and held us hostage inside that night.

In the morning, we were allowed to go outside. Robert and I roamed the streets, meeting up with other neighborhood friends. The world was transformed. Yards were unrecognizable; the skyline had changed. We crawled over downed pine trees, ignoring our mothers' pleas to stay out of the gleaming yellow sap which inevitably ended up in our hair. Yet even the most fearless of the pack, my brother, steered far clear of the whipping power lines.

The smell was heavenly, that sharp, fresh scent of pine. Chain saws buzzed, men hollered, and the air tingled with excitement. And then something miraculous happened: a team of mules came clanging through the fallen trees. Their hoofbeats rang like church bells; their leather livery rattled and squeaked with every step. This mule team had arrived on Brookside Drive to clear the debris, to make way for the trucks that could not yet get past the piles of trees.

For two days I hovered by the mule team, enthralled with the animals themselves who patiently let me pet them and rub my cheek against their enormous muzzles, and the men who drove the team with their mysterious commands and grunts. They were all larger than life, and nobody paid me any mind.

Looking back, it's hard to believe this really happened. We lived in a world of men on the moon, not mule teams. It wasn't even legal to have a horse in the city limits. And even then, in the days of children and dogs running free and untethered, the idea of my very small self, squatting down within inches of a working mule's hoof as big as a dinner plate, seems unlikely. Yet there they are, memories of the mule team standing out as a highlight in a magical childhood.

With every tornado since, I've revisited that memory. Now, fully adult and keen to the dangers, terrors, and tragedies tornadoes bring, I grapple with the combination of anguish for loss and the thrill of my pine-scented mule team memories.

Morning Musings on Bears
January 26, 2023

"I wouldn't mind being mauled to death by a bear," said our bonus-son Joseph over coffee recently, "as long as no one kills the bear."

I reached for my green notebook and jotted, "Important: If Joseph goes missing in the woods, do not report it!"

"I agree," my husband Joe piped in from the kitchen. "Could be worse."

"Do NOT go hiking with Joseph and Joe in bear territory," I added in my green notebook, underlining it three times.

Unlike Joseph and Joe, I do not want to be mauled to death by a bear and have lived my life accordingly. While I've camped in Africa where the only thing between me and the lions was a thin canvas tent, I have not camped where the grizzlies roam, nor will I.

Still, years ago I took our youngest daughter Anna on a vacation to the NC mountains. We planned hike to every waterfall in the vicinity. At the last minute I realized we'd be in bear territory during cub season. I called my father for advice, assuming he'd say, "Don't worry, that's black bear territory. They'd never hurt a soul." Instead he said, "I'll lend you my book, *How to Survive a Bear Attack*."

Armed with the book that would save our lives, I started to read. "If a bear tries to bluff, slowly walk away while speaking calmly to the bear. If a black bear attacks, fight back with everything you have! If a grizzly or brown bear attacks, play dead! If the attack persists, fight back with everything you have! If you surprise a bear of any species, don't fight, but if it attacks, see above." The rules went on, with conflicting instructions for survival totally dependent on a jumble of variables.

Despite the unlikelihood that I'd be able to call up this bear attack survival flow chart, I convinced myself that we had nothing to worry about and off we went. Sadly, we didn't see a single bear, and the creepiest feeling I got on our week-long hiking excursion was a sense of panic at realizing we were alone on a state park trail, where humans with bad intentions seem to lurk.

Were we safe from bears? Black bears (*Ursus americanus*) may be the continent's smallest bear, but they're still pretty big, with typical male adults weighing up to 550 pounds. They're fast, they climb trees, they can open jars and doors, and they are becoming increasingly habituated to humans. Still, with just 5 black bear attacks in the 2020s, I probably should not fret. Cars, however, our method of reaching this mountainous destination, take out around 35,000 people a year in the US.

If my fellas want to go out in blazing bear attacks, they're better off out west. Grizzly bears (*Ursus arctos horribillis*) are a species of brown bears, as are Kodiak bears, their larger cousins who weigh in at up to 1,500 pounds. Despite a size advantage, Kodiak bears are slightly less aggressive than grizzlies. For those running through the bear attack survival flowchart, you will know the brown bears by their prominent shoulder humps and ridges over the forehead, giving them a concave profile. Death by grizzly, the most aggressive bear, is still looking slim with just eight attacks in the 2020s.

So it's north to Alaska, home of the polar bear. Polar bears are the only animals who actively hunt humans, besides humans. Polar bears can smell their prey from 20 miles away, and if you want to scare yourself silly, know that they cannot be detected with night goggles. With the polar bears' habitat in decline, encounters with humans are increasing. Just last month, the first fatal polar bear attack in 30 years occurred in Wale, Alaska.

Polar bears may be mighty and fierce, but they are no match for climate change. No one can say for sure what's in store for these incredible creatures, but we do know that they are dependent on sea ice for survival. As their name *Ursus maritimus* suggests, polar bears are classified as ocean mammals, living on land only in warmer months when ice melts. Land life is a struggle for polar bears as the available diet is low in energy, and they are not well adapted to finding and catching it.

Who knows? Perhaps polar bears will adapt and return to the land. Evolutionarily speaking, they're still young, splitting from brown bears only about 500,000 years ago. Grizzly-polar bear hybrids are on the rise as polar bears wander inland, ahead of shrinking ice. Oh, polar bears, you were nice while you lasted, our future generations may think, staring at images, or maybe holograms, on as of now unimaginable devices.

Despite an uncanny fear of bears, I admire them, and respect them, and as with most critters am enchanted by them. And in truth, what's to fear? There are no bears as destructive as humans, who are, ironically, also capable of such glorious goodness.

Groundhogs, Love, and Fate
February 2, 2023

I was fresh out of college, a January graduate, and ready to change the world. While awaiting that perfect, if ill-timed, teaching position to open, I substituted in the Boston City schools. It was Groundhog Day, and I was assigned a 2nd grade class. 2nd graders love Groundhog Day, right?

"Guess what today is?" I asked the class, pulling a poster of a groundhog out of my Mary Poppins-esque teacher bag.

"F*** groundhogs!" called a tiny boy, using the real word. This was followed by a harrowing recess on the ice-covered playground during which that same fella climbed over the fence and skated away down a parking lot.

Never again, I vowed, and I began another job search. This being before the internet, job searching included driving around looking for potential prospects. I passed a sign reading, "Perkins School for the Blind." My heart pitter-pattered. With no experience in the field of deaf-blind education, I turned around and applied for a job. I was eager to learn and work with children whose disabilities eclipsed any I'd ever seen in mainstream public schools.

Unbeknownst to me, on that same day a handsome lad named Joe also applied to work at Perkins. We each began our new jobs on March 1, clueless as to what the fates had in store.

This summer, that handsome lad and I will celebrate our 36th wedding anniversary.

You'd think we'd have ceramic groundhogs decorating our shelves, or that we'd snuggle up to watch the famous movie every February 2nd. In fact, we just saw "Groundhog Day" for the first time this year. Other than sparking fate, it seemed groundhogs would play an insignificant role in our lives.

But lo! Our groundhogless days ended when Joseph (the bonus-son who wouldn't mind being mauled to death by a bear) set up the tipi in the yard. It attracted curious onlookers of all sorts from box turtles to rat snakes to possums and even a groundhog. Finally! A critter we'd never spotted in the yard. We watched it snuffling around in the kudzu on the creekside and spotted its footprints in the sand in the culvert under the road. That little whistle pig could disappear in a hole in the neighbor's yard and end up on our creekbank.

What an adorable animal! It was thick and round with tiny ears, standing up on its hind legs scouting out its territory, wary of us. And such a voracious eater! I had hopes it'd thin the kudzu, but no such luck. We loved our new friend and reported our sightings regularly. But it got me to thinking. I really didn't know what a groundhog was.

"It's a big rodent," one friend said.

"A marmot, right?" said another.

"A whistle pig?"

"A woodchuck?"

"Aren't they squirrels?"

Obviously, no one knows what the heck a groundhog is other than Kingdom: Animal; Phylum: Chordate; Class: Mammal. For those needing the refresher, remember Keep Ponds Clean Or Frogs Get Sick (Kingdom, Phylum, Class, Order, Family, Genus, Species).

Moving along the classification scale, groundhogs are indeed rodents, an order characterized by eternally growing incisors, and who make up about 40% of all mammal species.

Groundhogs split from beavers, mice, and other rodents at the family level, where they remain with tree squirrels, flying squirrels, and other ground squirrels in the family Sciuridae. Are they squirrels? Technically, yes.

So, what is a marmot? Marmot is a genus of the family Sciuridae, comprised of 15 species commonly called ground squirrels. I'll be honest, I've never heard of most of these marmots, but am intrigued by one called the Tibetan Snow Pig. Marmots are the largest members of the squirrel family, and one of few mammals that truly hibernate. Even bears don't actually hibernate, but instead enter a state called torpor meaning they can be awakened if disturbed. Beware.

The species name of the groundhog is *Marmot monax*. Common names include woodchuck, whistle pig, land beaver, and more. Baby groundhogs are called chucklings, the cuteness of which turns me inside out.

Now that I know what a groundhog is, I wonder why I've only seen one in the yard. According to Outdoor Alabama, they generally don't venture south of the 33rd parallel here in Alabama. Our lot is sitting comfortably at 32.6338317 degrees. This may explain their scarcity here, 0.4 degrees below their range. Mighty nit-picky, these groundhogs.

I'm hoping beyond hope that the groundhog returns, but with predators such as dogs, coyotes, and foxes, plus killer cars on the roads, their typical life span is only three years. A woodchuck seeking a home here is up against the odds. Despite our insistence that our little lot stay wild, we are surrounded by development which sends our sheltering wildlife elsewhere and our souls into a place of intangible longing for nature.

Happy Groundhog Day to all who celebrate love and wild things, and to those who make way for the least of these.

Cupid's Chemistry
February 9, 2023

St. Valentine's Day is nearly here! What better time to talk about *Faith, Madness, and Spontaneous Human Combustion.*

This collection of essays, subtitled, *What Immunology Can Teach Us About Self-Perception* by Gerald N. Callahan got me thinking about my immune system in a whole new light, including how my thymus, lymph, and spleen have influenced me in all ways romantic.

According to a research team at Rutgers, the hormones governing romantic love act upon us resulting in the distinct categories of lust, attraction, and attachment. Let's skip lust and go straight to attraction.

Mythology leads us to believe that we become attracted to one another when we're shot with Cupid's arrow. Science explains otherwise.

As with most things human, we are less in control of our emotions than we think. How we laugh at a peacock strutting his feathers; how we marvel at the brilliant crimson plumage of a male cardinal. We recognize those behavioral and physical adaptations geared for reproductive success in other animals, but when it comes to ourselves, we assume it's a matter of will, of personal choice. It is to a degree, but there are deeper tugs and pulls we are not aware of.

We call it chemistry. If we're feeling scientific, we call it pheromones, which are subtle chemical cues that act on us in mysterious ways. Whatever we call it, it's that je ne sais quoi that draws us nearer to, or repels us further from, another person as if Cupid's arrows were indeed involved.

My curiosity about attraction was first piqued by a passage in *Faith, Love, and Spontaneous Human Combustion.*

A quick aside, I am on my third copy of this book. The first was given to me by my dear friend Marian Carcache who is a true devotee. She knew I'd love it. Thank you, Marian. I have now passed on that copy and my replacement but am holding fast to my current copy due to my excessive marginalia.

It was in this book where I first learned that we are attracted to people with different immune systems. Bear with me while I get technical. There's a cluster of genes called the major histocompatibility complex (MHC) located on human chromosome 6. A cache of important information is located within the MHC, coding for proteins which recognize "self" and "non-self." In other words, they control our immunities: what will kill us, what we will survive.

When transplanting organs, matching MHCs are vital in selecting a donor. For the robust health of the species, however, different MHCs are the best when selecting a mate. The more varied our immune systems are, the more immunities we have. This is just one aspect of "hybrid vigor," making mutts healthier than purebreds, a biological basis for mixing it up and not mating with our cousins.

But how would we know who has different immunities? After finding that mice prefer to mate with mice who are MHC-different based on the smell of their urine, researchers turned to humans in what is now called the "Sweaty T-shirt Experiment." 44 men were each given a clean t-shirt to wear for two nights. 49 women were then asked to rate the t-shirts in terms of intensity, pleasantness, and sexiness. They consistently rated the t-shirts of MHC-different men higher in all categories.

This experiment has been replicated under different conditions and in varying populations, including a tight-knit religious group in South Dakota who marry amongst themselves. Even in that situation, married women have managed to select men with greater MHC differences.

Let's do our own experiment. Think of someone you've been attracted to. Have you ever picked up their shirt from a pile of dirty clothes, buried your face in it, and inhaled deeply? Felt bliss as that warmth emanates through your nose to your belly? You don't have to answer in public, but I see you.

Attraction, as exciting as it is, is dangerous territory. During this phase of romance the hormones dopamine and norepinephrine flood our brains, leading to giddiness, loss of appetite, and euphoria. Conversely, serotonin, the hormone involved in stabilizing mood and appetite, is suppressed, contributing to anxiety. It may be fun for a while, but our bodies interpret this state as one of intense stress and work to correct it.

Thus, the attraction stage is untenable. After a few months it gives way to attachment, or what I call the "pair of old shoes" phase. The hormones oxytocin and vasopressin take over as the other hormones settle down. These new hormones are responsible for bonding and contribute to stable societal structures like monogamy and family loyalty, all advantageous to the survival of the species. FYI, cuddling, petting dogs, and eating chocolate also cause the release of oxytocin, nicknamed the "love hormone."

Happy St. Valentine's Day to all you scientists, romantics, old-shoes, sweethearts, and tree-huggers! Whether you share this day with a romantic partner or not, may your oxytocin levels be a little bit higher on this holiday dedicated to love.

As for spontaneous human combustion, we'll save that for another day.

World Pangolin Day
February 16, 2023

Move over groundhogs! World Pangolin Day is just around the corner!

I confess that until a few years ago I was unaware of the plight of this precious animal. My curiosity was piqued, however, when a student told me that people eat pangolin scales.

"That sounds awful," I said, recalling a pangolin to be something like an armadillo, making a note to learn more about pangolins immediately.

Down the pangolin burrow I fell, and in no time became enchanted with pangolins, the planet's only scaled mammal, and the bearer of the horrific distinction as the world's most trafficked animal.

What are pangolins? Although they look like a cross between an anteater and an armadillo, they're more closely related to dogs and cats than to either of their look-alikes. There are eight species of these curious critters, the sole animals in the order Pholidota.

Pangolins live in Africa and Asia, making homes in trees or in burrows. Their front feet are burdened by enormous claws, curved and sharp for slashing into termite mounds, their heads sharp and pointy to snuff out their tasty food. One pangolin can eat up to 70 million insects a year, making them valuable members of their ecosystem.

Notably, pangolins are covered in hard, overlapping scales comprised of keratin, the same protein that makes up hair, fingernails, and rhino horns. When startled or scared, pangolins roll into a ball, their scales protecting the only vulnerable parts of their bodies: bellies and throats. The scales provide an excellent defense against leopards, lions, and other predators, but not against humans. The same adaptation that has protected them for 80 million years makes them an easy target for poachers who can simply pick up the frightened animals and toss them into bags.

Why would anyone pick up a frightened pangolin and toss it into a bag? Pangolins are a lucrative link in the black market.

In parts of Africa and Asia, pangolin meat is considered a delicacy, prized for its status as an extravagance rather than for its taste.

Additionally, until 2020, pangolin scales were listed in the catalog of Traditional Chinese Medicine, purported to treat everything from lactation difficulties to arthritis, although there is no research to support this. You might as well eat your fingernails.

Lest we begin shaking our righteous western heads at the practices of those from other continents, be aware that the thirst for pangolin leather cowboy boots, wallets, and belts in the U.S. has contributed largely to their declining numbers. Despite bans and laws, the black-market trade thrives, and the number of pangolins on our planet diminishes daily. All eight species are threatened or endangered.

Pangolins remain the most trafficked animal on Earth despite an international ban on the sale and consumption of pangolins.

Humanity and decency aside, other reasons compel us to leave pangolins alone to live their natural lives. In studying the pangolin trade to research the origins of the SARS-CoV-2 Coronavirus, scientists concluded that "moving wildlife species out of their natural habitats and into human dominated landscapes and large urban centers poses a serious and increasing risk of initiating epidemics from emergent pathogens in human populations." (Frontiers in Public Health, March, 2022.)

I have never seen a pangolin. Odds are a million to one that I will ever see one in the wild. Nothing about a pangolin has anything to do with me or mine except the thread which connects us all. Yet not a day has passed since I learned about these mystical, snuffling critters that I haven't thought about them, minding their business, poking about, mama pangolins carrying their pangopups on their tails.

I've also thought about people who protect pangolins. I've learned of organizations whose members rehabilitate rescued pangolins and see them through to their return to the wild, tracking them once released and even hiring airplanes to assist in locating them when they go missing. I've pondered the differences between billionaire fat-cats who pay over $600 per kg for pangolin meat and volunteers who spend their sleeping hours keeping watch over rehabilitated pangolins. I've extended compassion for poachers who risk their lives in illegal operations in order to scrap together a living in an impoverished land, and I've placed blame on systems of oppression that utilize humans and other animals as tools to line the coffers of the extremely wealthy.

Through a class I teach online and my use of a stuffed animal pangolin I purchased from World Wildlife Fund as my teaching companion, I've introduced hundreds upon hundreds of children around the country to pangolins, eliciting exclamations of love for these quietly lumbering beasts. My hope is that this love will extend to all persecuted beings and that someday, perhaps when my granddaughters are my age, we will look back in disgust at how humans once treated animals, including pangolins and each other.

World Pangolin Day occurs annually on the third Saturday in February. I encourage all of you to spend a bit of time to learn more about these incredible animals. Happy World Pangolin Day!

Mammals and Viruses: Connected through Time
March 2, 2023

It's beautiful outside, and I'm sitting on the porch reflecting on the existence of mammals. Today's mammalian musing is brought to me by the SARS-CoV-2 virus, which has finally caught up with me resulting in my having developed COVID-19.

As some might say, I've caught the coronavirus. Well, one of them. Coronaviruses, discovered in the 1960s, are one of many virus groups.

Seven coronaviruses affect humans. Four of these with boring number-names cause mild to moderate upper respiratory problems, some resulting in common colds. The remaining three are more sinister, so menacing that you've likely heard of them: SARS, MERS, and SARS-Co-V-2 which has been headlining like a Super-Bowl half-time performer since 2020.

If you want to be considered too big for your britches, you can go around correcting those who say, "I've got the coronavirus," to "You mean you've got COVID-19, caused by the SARS-CoV-2 virus, a type of coronavirus," but if you do, folks will turn your picture to the wall.

As I was saying, I have COVID-19, as does our daughter Emma. Joe remains unaffected; knock on wood.

Too sick to read or write on the day I tested positive, I lay in bed trying to visualize my cells. I imagined my healthy cells being coerced to produce these brand-new viruses, and I sent all manner of good vibes to my immune system, to my killer cells who might recognize the specific coronavirus spike and set out to destroy it. I welcomed new killer cells into the fold of fighters who came into being when faced with synthesized, look-alike spikes as a result of my vaccinations and boosters, cheering on my immunity with no regard as to how it got there. However the body is exposed, through sneezes or needles, immunity is its natural response. It's all "natural immunity."

When my fever ran high, I imagined the virus being baked into oblivion, and the sweat that poured out of my skin the product of the hard work my body was doing to heal. The voice of my grandfather Douglass scolded, "Sweat it out, sweat it out," every time I reached for the Tylenol. I tried not to think about this particular virus's proclivity to attack lung cells and was grateful for the advances in vaccinations and treatments thus far.

And now, on day three, I'm exhausted but sitting upright, ready to chat about viruses.

Viruses are some bizarre little things. What in the world is a virus anyway?

"It's a bit of genetic information wrapped up in a protein coat," say the students who took my online class, "Keeping Up with the Kingdoms."

Viruses can't reproduce on their own. They multiply by hijacking the machinery in healthy cells, inserting their genetic material to replace the original instruction manual, and transforming these healthy cells into virus factories. Viruses are the most plentiful and diverse biological entities on the planet. Notice the precise use of words. They aren't living things, but because they have that bit of DNA or RNA, "genetic material," we can use the word "biological" to describe them.

There are so many viruses. One group of viruses, called retroviruses, embed their genetic material in our own DNA, which is a whole 'nother level of insidiousness. HIV is a retrovirus, which is why it never leaves once it takes up residence.

There are other retroviruses which have been with us since the dawn of mankind, so deeply embedded that they now make up 8% of our genetic code. These viruses, called Human Endogenous Retroviruses (HERVs) have significantly affected our species, providing embryonic immunity and the ability for our embryonic cells to develop into specialized cells, thus making us humans.

Researchers are studying HERVs and possible pathways to understanding MS and ALS, with hopes of better treatments and even cures for these terrible diseases.

But what of these mammalian musings I spoke of earlier? As you know, we mammals give birth to live young as one of our distinguishing characteristics. Without an eggshell to separate a growing body from its mother, it's a wonder the immune system of the pregnant mother doesn't destroy this foreign being in her body. This is, in fact, exactly what would happen if the mammalian mother and the developing fetus ever came into contact. Enter the placenta, the body's only temporary organ and a wonder to behold. The placenta has the hugely important job of connecting mothers to their unborn offspring while keeping them separate at the same time.

The sticky substance that binds the placenta to the uterine wall is comprised of a protein whose DNA is almost identical to a specific HERV, suggesting that it was the presence of an ancient retrovirus that eventually allowed for the mammalian ability to bear live young.

Oh, my stars, the body is in incredible, awe-inspiring ecosystem. The more we understand it, the safer we are from the dangers that once killed us willy-nilly. Every day I am beholden to the researchers who obsess on our tiny worlds within. This week I am especially grateful.

Drink your water, wash your hands, and be well!

I Survived the Midway Roller Rink
March 9, 2023

I was recently thumbing through a scrapbook I kept in high school when I found a treasure I'd overlooked a thousand times: my membership card to the Midway Roller Rink. It's poked through with thumbtack holes, meaning it had at one time lived on my bulletin board. Relics from the bulletin board ended up in scrap books, making way for more relics, seeing to it that I have a lifetime of memories.

As it was meant to do, this little yellow card took me all the way back to elementary school and the thrill of days spent at the Midway Roller Rink. The Roller Rink was out on Opelika Road, somewhere near the Auburn/Opelika border I'm guessing given its name. Housed in a long, narrow building, the Roller Rink was noted for its wooden skate floor. I can hear and feel the clackity-clack of the boards under my wheels as I navigated the long, narrow rink. There were gaps, loose boards, and all manner of snares to foil a young skater, including splinters that could end up in the rump when you fell.

I, like most children in town, learned to roller skate at the Roller Rink with the help of the wooden handrail which ran the perimeter of the skate floor, clutching it in terror as both feet wheeled out from under me over and over. Dragging myself along amidst the clang and crash of the advanced skaters whizzing by, dazzled by the disco light flashing in time to "Love Will Keep Us Together," I did okay until I reached the Gaping Maw, the Dark Void, the Abyssal Pit of Hell otherwise known as the Boys' Bathroom.

The gateway to the Boy's Bathroom was an open recess in one of the long walls of the rink. How the boys ever got in and out of there alive I still don't know. There the handrail ended, and I had to get past the black rectangular hole in the wall without falling into the pit, which was about a foot lower than the floor of the rink. I'd reach the gap and stare longingly across the expanse, which was probably about three feet, to its other end, and pray that Guy Montgomery or Matt Samford wouldn't come flying up behind me and push me in while I straddled that vulnerable no-man's land.

What if I didn't make it? What if my wheels stopped turning and I was stuck there, framed by the evil mouth of the boys' room, unable to move backward or forward, for all eternity?

There were no comforting parents to hold our hands and guide us through this and other land mines at the Roller Rink. They would drop us off there and "run errands," which I now think consisted of cocktails and card games.

I eventually learned to skate well enough to make it around the rink without holding on. I joined the cool kids who would stand in front of the larger-than-life fan that was built into the wall and see if it would blow us backwards. When old Mr. Crawford's perpetually grumpy voice would come over the microphone and announce, "Reverse Skate," I could change direction without falling down and having my fingers crushed to bits under someone's wheels.

Best of all, I could play Crack-the-Whip, the same version of the neighborhood game but on wheels, which was terrifying and thrilling. I preferred to be somewhere in the middle of the human chain. The person in the center had to be able to pivot on wheels, and the person on the end, well, they got "cracked" by the serpentine energy that traveled through our arms. I am certain broken bones were involved.

My parents always gave me a quarter to spend at the Roller Rink, which was just enough for a pack of ancient "Now and Laters." It's a wonder my teeth didn't fall out trying to chew those interminably sticky squares of fruit-flavored sugar. I could make my candy last a long time, much to the chagrin of Mrs. Watkins who glared at us when we hung out in the tiny lobby. She and Mr. Watkins were Old-Testament Old and quick to reprimand. We were all afraid of Mr. and Mrs.

Watkins, which was likely ridiculous, but it made the Midway Roller Rink more exciting.

By the time my own children were learning to skate, the Midway Roller Rink was long gone. We had Battle's Skate Center, which was bigger, flashier, and not a place you'd dream of dropping your elementary-school-aged children off while you went out for "errands."

I longed for the Midway Roller Rink then, and still do.

It's funny how each generation wishes the next could experience the same things they did, even if those experiences involved a skating rink run by a frightening and ancient couple, which featured a fearsome cave of horrors, splinters in the hind end, and inevitable broken bones.

North Brother Island
March 16, 2023

About six months ago our youngest chickadee, Anna, flew far from the coop, all the way to New York City. Last week, I finally visited her in her Brooklyn home. She and two other Alabama transplants live squeezed into a narrow apartment, rooms stacked on top of each other, along with two dogs and a cat. They move around like an eddy, always circling, never bumping. In no time, Anna has become adept at navigating not just her living quarters but a city bigger than I remembered, bigger than I can comprehend.

Anna and I walked miles and miles in those three short days, taking in museums, gardens, restaurants, and so many train rides. These days you don't get your money out; you just tap your phone on a magic square, tap-tap-tap, and the currency flows from your bank account, whoosh-whoosh-whoosh, and into the hands of the vendors. Doors open, food appears. Tap-tap-tap.

What a marvel it is. How do so many people live so close together? Every inch of space is utilized, yet, in an enormous gift of grace, the city carves out places for people to sit and gaze at water, gardens, and sky. Crocuses bloom in abandoned railroad tracks on the High Line. Grasses recover and twigs burst with buds on the Little Island. On every corner, markets explode in color with bouquets of flowers, and the humans, displaying every hue of the melatonin rainbow, stop and tap their phones, gathering bundles of blossoms to brighten up their own narrow apartments.

Even in a megacity, nature finds a way.

When I got home, my brain got to wandering and puzzling over what would happen if all the humans vanished from New York

City. This took me down a tunnel in which I stumbled upon a place called North Brother Island. How is it I'd never heard of this abandoned bit of land?

North Brother Island and its companion, South Brother Island, are located in the East River between South Bronx and Queens, half a mile from Rikers Island. This island is not in some remote location but right in the bustle of the city. You could paddle a kayak from the Bronx to the island, but it would be dangerous and illegal.

North Brother Island has a ghoulish history. Due to treacherous currents, it lay undeveloped but for a lighthouse until 1885, when the Riverside Hospital built a unit there to treat and quarantine victims of smallpox and other diseases including typhoid, tuberculosis, and polio.

In 1905, the Steamboat George Slocum caught fire and sank nearby, killing 1021 passengers. Dozens of dead bodies washed ashore. Three years later, Mary Mallon arrived on the island. Known as Typhoid Mary, she was confined to North Brother Island for over 20 years, eventually dying there.

By the 1930s, the quarantine hospital was no longer necessary. The island briefly housed WWII veterans and their families but largely lay untouched until the 1950s. At that time a drug rehabilitation center for adolescents was set up on the island. Corruption, cruelty, and costs forced this facility to close in the mid-1960s and the island has remained abandoned ever since.

Now given over to nature, the buildings, including the four-story tuberculosis hospital, the morgue, and the coal-fired power plant, are overtaken by trees and vines. Walls have crumbled, roofs have caved in, windows have shattered. Drone images show a lush green patch of land in a concrete sea.

Today both the North and South Brother Islands are protected wildlife sanctuaries, providing a necessary resting and nesting spot for herons and other colonial wading birds. The islands are managed by NYC's Parks and Recreation Department and supported by the National Oceanic and Atmospheric Administration (NOAA), the

Coastal and Estuarine Land Conservation Program, and the Trust for Public Land.

"I want to go!" Anna said when I told her about the abandoned wildlife sanctuary. So do I, but unless we can prove we have a scientific or academic mission that will benefit the health of the island, we are forbidden. Researchers can apply to visit the island during non-nesting months, and if granted access, must charter their own boat and remain accompanied by a NYC Parks and Rec staff member.

Although there are some who foolishly brave the currents, laws, and dangers, this sneaking onto the island for dares and adventures is ill-advised. If a building collapses around you or you should fall into a crevice of doom, no tap-tap-tapping of your phone will save you. No doors will open, no food will appear.

You and the ghosts of Little Brother Island will be entirely alone but for the gulls, herons, cormorants, and egrets who benefit greatly from this treacherous, uninhabited sanctuary.

Again, I am awed by and grateful for the people who devote their energy to saving wild spaces, including those in the most unlikely spots.

A Curious Colony
March 23, 2023

Spring is official now. The azaleas are strutting their stuff, the native buckeyes are dotting our messy yard red, and the college kids are hitting the Alabama beaches. Some have returned with reports of big purple jellyfish. "They looked like plastic bags!" one of my college friends exclaimed, eyes wide. "They were everywhere!"

I'd recently read reports of Portuguese Man o'Wars washing up on Panama City Beach and nearby areas by the thousands. Earlier this year they decorated the shores of the South Carolina barrier island we frequent, and we found a few ourselves while visiting over Christmas.

All this talk of these fearsome sea critters took me back to a childhood visit to Orange Beach, when my brother Robert was stung by a Portuguese Man o'War. I don't remember what my parents did to treat Robert's stings; I suspect they took him inland to a doctor. I couldn't be bothered. Robert was always getting hurt. If there was a child within a thousand miles who was going to get stung by a dangerous jellyfish, it'd be Robert. Furthermore, it's likely that he was hurt when picking up the washed-up jellyfish with the intention to throw it at me.

We returned from that trip with several specimens of sea creatures, as usual. This time the collection included two Portuguese Man o'Wars: one floating in a jar of formaldehyde, the other dried up like a hardened balloon, no tentacles, just a bloated, empty body, like a hardshell balloon. Don't ask me how we happened to have formaldehyde with us on a family vacation. It's just one more childhood mystery.

I really liked that dried up Portuguese Man o'War. It smelled like the ocean. Even though it had gone pale, you could still see the purples and pinks, especially when you held it up to the window. The shape and color of this balloon-like part of the animal supposedly resembles an 18th century Portuguese warship, giving the creature its name, but I knew nothing of that.

My up-close experiences with this critter had me wrongly thinking that I knew something about Portuguese Man o'Wars. "Dangerous jellyfish," I'd say assuredly. "Pretty animals, all purple and pink with long, crinkly tentacles."

It turns out a Portuguese Man o'War is not a jellyfish. Like a jellyfish, it's a predatory animal that uses stinging tentacles for stunning its prey as well as for defense, drifting along where the wind and ocean currents take it. A jellyfish, however, is a single organism, while a Portuguese Man o'War is a siphonophore, an animal made up of lots of small organisms living attached to each other in a colony.

Excuse me?

Each Portuguese Man o'War is comprised of a group of zooids, or individual organisms that live together with other zooids in a colony. Each zooid has a specific job to do, so specialized that they cannot live alone, and are named appropriately. In the case of the Portuguese Man o'War, one zooid, the pneumatophore, is the purple and pink balloon part, called the float. Others are gastrozooids which are the feeding tentacles, dactylozooids which make up the food-capturing and the defensive tentacles, and gonozooids specializing in reproduction.

There are 175 recognized species of siphonophores encompassing a wide range of characteristics. All are ocean critters, and most exhibit bioluminescence. The Portuguese Man o'War was the first siphonophore to be described by our taxonomy friend Carl Linnaeus. About three years ago, a giant siphonophore measuring almost 50 feet long was discovered off the coast of Western Australia, making it the largest species on record. Many siphonophores are delicate, breaking apart if touched.

As Robert Louis Stevenson wrote, "The world is so full of a number of things, I'm sure we should all be as happy as kings." To learn that a Portuguese Man o'War is made up of a colony of zooids is just what I needed to discover today. Crown me king!

Happy spring to all. May your week be filled with wonder and awe.

Out on Highway 61
March 30, 2023

At the onset of a most ordinary evening back in November 2016, my phone rang. I smiled on seeing that it was Emma, our middle chickadee, calling to check in during her road trip from New Orleans to Little Rock.

"Mom? My car broke down and I'm in a field in the middle of nowhere on Highway 61. Hang on. There's a tornado warning." I hung on. She hung up.

I called back frantically, relieved when she finally picked up. A truck had stopped to help. I heard the voices of men, big scary men, in the background and my mind raced. If I called 911, they could find her, right? But among those in the truck was the mayor of the tiny town where she'd landed. Surely she was safe.

While I stood shaking 400 miles away, my mama-bear brain running ninety-to-nothing even with the presence of a mayor, Emma was grabbing her viola out of her broken-down car and hopping into a truck with total strangers who were taking her to the local motel.

The Bob Dylan lyrics ran through my head, "Abe said, where you want this killing done? God said, out on Highway 61."

"I'll call you back, Mom," she said over and over, hanging up on my desperate calls. I wanted to keep her on the line, that mysterious connection of satellite signals fueling a false sense of security.

"Everything's fine," she assured me in the morning. Her car had been towed to the mechanic, and as soon as she'd ventured out into the light of day, she'd been embraced by the town folk, taken to a church, fed, and even housed. Now she was leaving the motel for a mother-in-law cottage in a local family's home.

Three days later, I was on the phone with Melvin, the mechanic who fixed her car but at a sacrifice to the catalytic converter.

"It'll get her from point A to point B," he said, "but she can forget about point C."

There in my notebook on December 2, 2016, the words, "GET EMMA HOME!" are followed by, "Emma is on the way home," and finally, "Emma is home," a satisfying resolution.

This adventure marked a turning point in Emma's life, a clarifying realization that she was living a bit chaotically. It gave personal weight to the story of Robert Johnson at the crossroads on the same Blues Highway.

Emma regaled us with tales of the people she met, their personalities, their quirks, their smiles, their generosity. Her story moved permanently into our familial lore with a glowing warmth in our hearts for that town we'd never heard of before then.

That town was Rolling Fork, Mississippi.

This week, news of Rolling Fork hit the international circuit when it was flattened by a freakish EF-4 "wedge tornado." The family that took her in no longer has a house or a mother-in-law cottage to offer a stranger with a broken-down car.

It's all too familiar, isn't it? It seems like yesterday when our own local community of Beauregard was in the same spotlight as Rolling Fork. Pick a day, any day, and your heart will shatter over loss and catastrophe from floods to wars to diseases to wildfires, on and on.

None of us can respond to every crisis. Those compelled to action toward any particular disaster can choose from a vast selection of relief efforts, depending on their resources. Some folks have large wallets, some folks have large trucks and chainsaws. I have neither of these, but I've got a powerful tool for this disaster: story.

Throughout human history, story has connected us to each other's humanity. Story persists, story builds, story reminds us over and over that we are all wayfaring strangers on our own hero's journey. As I tell the story of Emma, broken down and sheltered in Rolling Fork, Mississippi, I am reminded of the extraordinary wonder it is that each of us has a life, a history, a place. I am amazed that Emma's story, and now our family story, has this unlikely thread of Rolling Fork, Mississippi woven in. And having read this far, now you too have a story about the good folks in Rolling Fork.

(Over a year later, the town remains devastated, though some businesses have returned. Many residents returning to the town say it is still unrecognizable.)

Little Drunk Raccoons
April 6, 2023

My dad enjoyed a cigarette with his morning coffee and his
newspaper his whole life. Often, he met with a group of his peers, *The
Order of the Geezers*, for breakfast. For years they met at Hardee's on Gay
St. When I was in college in the early 1980s, I'd phone Hardee's if I
needed him before he went to his office.

"Dr. Mount, your daughter's on the phone!" I'd hear the busy
cashier call out to the diners. No one missed a beat.

As laws regarding smoking evolved, Daddy and the Geezers
had to change restaurants, running from the anti-cigarette coalition.
Eventually, Daddy moved his morning ritual to Fat Daddy's, a honky-
tonk pool hall and bar conveniently located outside the city limits and
therefore out of reach of that particular arm of the law. By that time,
The Order of the Geezers had dwindled, and there were times when Daddy
ate alone, pondering politics, society, and the changing world by
himself. Not that he minded; he was quite fond of his own company.
He fell comfortably into spending every morning at Fat Daddy's where
he smoked and ate and read the paper, enveloped in the musty cigarette
and beer vapors of the night before.

One morning he called me during his breakfast routine. "Mary,
come down here to Fat Daddy's. Bring your camera."

I arrived and was surprised to see that his was the only vehicle
in the parking lot. I pulled up beside the truck. He opened the door and
a thick cloud of Natural Indian Spirit cigarette smoke poured from the
cab. He led me behind the club to the Dempsey Dumpster, grinning
from ear to ear like a kid in a candy store. I knew that grin; it was the

one reserved for amusing critters, the same one he wore when he'd sprinkle ants onto the dogs' backs to watch them squirm.

"Look at those little drunk racoons," he said when we got close, gesturing with his cigarette to the inside of the very stinky dumpster.

Sure enough, three racoons had found their way into the Dempsey Dumpster and they were rolling around drunk as skunks in the dregs of the beer thrown out the night before.

Poor little drunk raccoons! They were stumbling over each other, wet and stinky and so confused! One little fella, maybe a little less drunk and a lot less wet than the others, stared at me plaintively and I got a clear, lovely shot of his face, which remarkably was not drenched in beer. Their shenanigans were funny for a minute, but it was clear they needed help.

We searched around the back of the parking lot and found some old two by fours. We turned them into ramps for the raccoons to crawl up and out of the dumpsters, which one of them did right away. The others were not yet sober enough to successfully maneuver the escape.

"Buy me breakfast?" I said after we'd had our fill of the little drunk racoons.

But alas, not only was Fat Daddy's closed, they were no longer serving breakfast a'tall, and hadn't in months.

"Wait a second," I said, puzzled. "Why are you here?"

It turns out that the man was not to be deterred from his routine of dining at Fat Daddy's, closed or not. Now he simply purchased a sausage biscuit and a coffee from the Hardee's drive-thru, drove over to the parking lot of Fat Daddy's, and there he sat in his truck with his cigarettes, newspapers, and meal, happy as a racoon in beer. Every morning. In the empty parking lot. In his truck.

We could stop now and ponder the little drunk racoons, and the proclivity of some wildlife to drink till drunk if given the chance, but I'm distracted by the proclivity for humans, especially men, to age into what I call the Routine Rut. I'd seen it before as one who worked in restaurants. The men would show up on the same day at the same time to the same table, and we'd have their signature cocktail on that table before they sat down. Thou shalt not vary the routine.

I miss my dad and his quirky routines and ways. Still, I'm keeping a close eye on my husband Joe, looking for signs of that Routine Rut. Every time he suggests we walk in a different direction on our daily dog walks, I breathe a sigh of relief, holding out hope for the powers of neuroplasticity.

For those worried about the little drunk raccoons, fear not. They were gone the next morning, which is more than I can say for my dad who continued to show up to the parking lot, day after day, year after year.

Unlearning Fears and Phobias
April 13, 2023

Our Alabama spring is past the glorious azalea and dogwood tree announcement phase and has now settled down to the business of securing a successful place in the world for our flora and fauna. Groundhogs are rushing to nourish their babies, who must mature and head into hibernation in just seven months. The pollinated flowers are busy turning into fruit with the goal of seed dispersal, and our snakes and turtles, who have been quietly brumating all winter, are out looking for food and mates.

Last week, while walking around a pond with our granddaughters Ruby and Annabelle, we saw several water snakes sunning themselves on the rocks. I encouraged the girls to walk quietly so as to get a better look, and showed them how to identify the snakes as non-venomous water snakes.

"Look for the up-and-down stripes on the jaw and the rounded eyes," I explained. "Those are harmless water snakes."

We didn't see any cottonmouths, also known as water moccasins, distinguishable from non-venomous water snakes by their "Zorro masks," the horizontal black bands across their eyes topped with notable ridges and the absence of vertical stripes on their jaws.

After just a few minutes, the girls got their snake eyes working and each of them discovered more water snakes.

"Oh, these sneaky-snakes are so cute!" they said, making squealing noises usually reserved for puppies.

We're a critter-loving family, but we're not careless. We taught our children, who in turn taught their children, to stay away from snakes they cannot identify, just as we taught them to chew their food well, and to buckle their seatbelts in cars.

Hearing our granddaughters talk about snakes so lovingly got me to thinking about learned behaviors and innate fears. I've heard that folks are born with a fear of snakes, but that doesn't go along with my lived experiences. Eyewitness and anecdotal evidence is weak, however, so I spent my morning digging into the research on this topic.

Oh, my ever-loving, bias-confirming stars. Let me tell you what I found.

A research article, published in 2022, entitled "Nature relatedness: A protective factor for snake and spider fears and phobias," studied the effects of urbanization on people's animal fears and phobias. In the introduction, the authors cited a growing body of research linking higher levels of happiness, increased emotional resilience, and a lower risk of psychiatric disorders to nature relatedness, defined as one's cognitive, affective, and experiential connection to the natural world.

The authors went on to explore a possible correlation between nature relatedness and snake and spider phobias. Their findings concluded, no surprises here, that incidences of these specific animal phobias are inversely correlated to nature relatedness.

This research, the results of which seem so obvious that it hardly needed conducting, turns out to be beneficial in two ways. First, it suggests that fostering our connection to nature might prevent irrational fears of spiders and snakes, and second, that people with specific animal phobias can be helped by education about and positive, incremental interactions with the animals they fear.

Past conclusions that fear of snakes is innate could be clouded by the myriad messages infants receive about snakes, including the tone of voice associated with snake conversations and the reactions of adults around them. Babies from different cultures have different reactions based on the prevalence of venomous snakes in their local regions, pointing further away from an innate fear.

From my morning ramblings in research, I have a few take-aways of my own. For one thing, infants are a lot smarter than we give them credit for. Tuned in to tones and suggestions we don't even know we are projecting, they are absorbing our messages and carrying them into adulthood. For another thing, every intuitive feeling I have about the importance of nurturing our connections to the natural world proves to be solid. We need to be outside, preferably with our children, a lot. And finally, there is hope for those who are burdened with irrational animal fears. I understand that these phobias can be debilitating, and I have sympathy for those suffering.

For anyone wanting to learn more about our local snakes, I suggest starting with *Outdoor Alabama*. On their website, you can find excellent descriptions of Alabama's six venomous snakes. Additionally, there is an excellent Facebook group called "Reptile and Amphibian ID & Education," where you can practice your identification skills, find folks to come remove venomous snakes from your premises, and enjoy basking in the company of people who proudly admit their transformation from a "The only good snake is a dead snake" mentality to that of appreciating these beneficial members of our ecosystem, even if from a distance.

And of course, there's the best teacher of all: nature. This is a great time to walk around in wonder as the sparse winter woods give way to abundance. The animals are out in force after a long, hungry winter, and if you're lucky, you'll find some wild things to remind you that we are not alone on this planet, that we are a part of an ecosystem that was here long before we arrived on the scene.

Hello, Granddaddy Longlegs
April 20, 2023

Aren't these phone cameras amazing? I often use mine like a
magnifying glass, taking a picture then zooming in for details. Today,
zooming in on the picture I'd just taken of a mountain laurel blossom,
I was tickled to find that it had been photobombed by a granddaddy
longlegs. This spindly critter hiding out on the pink petals was one of
the first creepy-crawlies I learned to identify as a young child.

"Hello, spider," I whispered, as we do to pictures of animals,
but then remembered those long afternoons with the neighborhood
gang, lounging around in giant pine straw nests we built to live in
because our parents didn't let us come inside the house until dinner
was ready. We'd while away the hours sharing all kinds of wisdom
gleaned from older kids, encyclopedias, and Ranger Rick Magazine.

"If you kiss your elbow you'll turn into a boy," my brother
insisted, but it could not be proven because we couldn't kiss our
elbows.

"If you put salt on a bird's tail you can catch it," Dana said, but
again, who can get close enough to a bird to put salt on its tail?

"Granddaddy longlegs are the most venomous spiders on
Earth, but their fangs are too small to bite with," Claude chimed in.

"Dummy. They're not spiders," answered my brother. Claude
hurled a pinecone at him, thus saving face.

And then we all grew up and stopped pondering elbows kisses,
salted tails, and granddaddy longlegs. That picture I saw today,
however, brought it all back to me.

"Hello, whatever you are," I amended.

If granddaddy longlegs aren't spiders, what are they? Both insects belong to the class Arachnida, thus are arachnids, but spiders and granddaddy longlegs diverged millions of years ago. Spiders are members of the order Araneae, whereas granddaddy longlegs are one of the Opiliones (Opilio means sheep-master in Latin), also known as harvestmen.

You can tell by looking at a granddaddy longleg's body that it's not a spider. We learned in elementary school that spider bodies have two prominent body parts: a cephalothorax and an abdomen. They look like they are wearing very tight belts. Granddaddy longlegs, on the other hand, have little button bodies with all parts fused.

Think back to your own experiences with these long-legged bugs. Likely you've never seen one in a web. That's because they don't spin webs. They don't have silk glands like their spider look-alikes. Furthermore, they only have one pair of eyes as opposed to eight.

In fact, granddaddy longlegs are more closely related to scorpions than spiders, a handy fact with which to impress children.

Someone told me, maybe Claude, that the best way to pick them up was by one of the long legs. I have been doing that all my life, but have now learned the error of my ways. The long pair of legs, second from the head, are specialized for capturing prey, smelling, and even breathing. While harvestmen can survive losing a leg, they cannot survive losing their long legs. They are known to self-amputate to fool predators, but they will never surrender a long leg. Remarkably, an amputated leg can twitch and tease for up to 30 minutes, giving the harvestman time to escape. Best not to handle them by the legs after all.

As for the venom, there is none, nor are there fangs. Unlike spiders, who use fangs to inject venom and turn their prey into goop, granddaddy longlegs can eat little chunks of food. They are one of only two species of arachnids that eat vegetation as well as fungi, snails, worms, bird droppings, aphids, and dead bugs. Excellent house cleaners, even pest control companies suggest letting them live amongst us, with gentle relocation outside as the suggested method of removal.

Ever the word nerd, I'm curious as to why they're called harvestmen. Some sources say it's because their extra-long pair of legs reminded folks of shepherds' crooks, while others say it's because they are abundant in harvest times. According to Frank Cowan, author of *Curious Facts in the History of Insects*, they are so named because of a belief that if you killed one, you were doomed to an unproductive harvest. A wealth of lore suggests that they bring good luck. One legend tells that they will even help you find your lost cattle by pointing the way with their long legs.

I should clarify that there is an actual spider that some people also call granddaddy longlegs. It has a clearly jointed body shaped like a slender peanut. These spiders are nicknamed cellar spiders or vibrating spiders. There is also a flying insect, the crane fly, with extremely long legs. Some people, no self-respecting person I've ever known, call that a granddaddy longlegs. We call them giant mosquitoes but know in fact they are not, and we refrain from swatting them.

One last thought: granddaddy longlegs or daddy longlegs? I've heard both and am now curious why some folks choose one over the other.

Ode to Insects and Wildflowers
April 27, 2023

Annabelle, our oldest granddaughter, and I were outside one morning last week wandering in the yard, looking for critters. After several minutes, I told Annabelle I had to go back inside because the little bugs were driving me crazy.

"That's because you're standing in the middle of Pollinator Paradise, Mimi," Annabelle said.

Sure enough, the fleabane, clover, spiderwort, and lyre leaf sage were up to my knees. Bees buzzed, butterflies flitted, and tiny, unidentified flying objects surrounded me.

"Why hello, pollinators," I whispered, my emotions righted. "You're not driving me crazy at all. Carry on."

I remember a time when the very wildflowers I just mentioned gave me anxiety. I've never been one to obsess over a tidy lawn, but still, seeing those brightly colored flowers taking over what little patch of sunlit space I have made me feel like an irresponsible homeowner. I'd whack them down with a weed eater and marvel that the blanket of broken green detritus almost looked like a wee grassy bit of yard.

It's funny how perception changes. Now, when I look out on my tiny Pollinator Paradise, I am calmed. I wander amongst the wildflowers, admiring the even petals of the friendly fleabane, the intricate anthers of the spiderwort splashing brilliant yellow against royal purple, the funny little down-turned trumpets of the lyre leaf sage. And oh, the clover, the puffballs and the perfect leaves!

It's the evenly mown lawns of grass that give me anxiety now. I'm worried about the bugs.

Researchers studying insect populations have seen a steady decline over the past few decades, but in the last five years, those numbers have plummeted. Round up the usual suspects (no pun intended): pesticides; light pollution; climate change; deforestation; urbanization; and the biggest, baddest, bogie man of all, intensive agriculture.

To a generation who played in the fog behind the bug spraying van as children, fewer bugs might sound like a good thing. And yet, so go the insects, so goes the planet. Insects provide us with at least $57 billion a year in ecosystem services like aerating lawns, dispersing seeds, disposing of dead organic material, recycling nutrients, and of course, pollination. But can we put a price on their value?

In the words of Francisco Sánchez-Bayo, of the University of Sydney, Australia, author of a recent analysis of the decline of insects, "If insect species losses cannot be halted, this will have catastrophic consequences for both the planet's ecosystems and for the survival of mankind."

Sweeping global reforms in how we eat, farm, and cohabitate with the natural world are needed to stop this cataclysmic scenario. Government protections and regulations are necessary if we are going to slow this mass extinction down.

So, what are we, individuals with limited power and resources, do to?

We can start by letting our lawns relax. No Mow May, the idea of letting your lawn go during this month, has been popular for years. Here, where our wildflowers bloom earlier, we should extend that to No Mow March, April, May. If you must mow, raising the blade on your lawnmower to allow grass flowers to bloom, leaving patches of weeds (aka wildflowers), and mowing your lawn in stages, no more than a third at a time, are easy practices to adopt.

For the flower garden enthusiasts, the insects will appreciate your replacing ornamentals with native flowering plants. Audubon (audobon.org) has a user-friendly native plant directory to help you select flowers and plants for your specific location.

Quite obviously, we can reduce or eliminate weed-and-feed applications and pesticides. Spraying lawns for mosquitos and weeds also kills the beneficial bugs and their food.

Or we can be lazy and let nature do the work, which is my preferred method of pollinator-friendly living. We rarely pull weeds, and when we do, we're selective. When the limbs fall, we leave them be. We'll whack at the invasives when we have the energy, carefully working around the natives.

"Perception is reality," my stepmother Janie used to tell me all the time. In my reality, an overgrown, tangled, messy yard is far more beautiful than a pristine lawn. For the sake of the insects, therefore the planet, I hope this reality catches on.

Roscoe, the Family Crow
May 4, 2023

One day when I was ten years old, my dad put on his tree-climbing boots and body harness, shimmied up a pine tree, and stole a baby crow. I'm sure he broke all kinds of laws doing this, and I imagine he had to fight the mother crow as well as former offspring and other crows who would have helped defend the nest from this horrid human predator.

Regardless, I imagine my dad did his plundering in a state of elation. He must have planned this thievery carefully, watching that nest by the hour through his binoculars for just the right moment. He'd always wanted a brand-new baby crow.

We took turns feeding the little rubbery bird, keeping it warm and watching it change day by day. As planned, the moment it opened its eyes my dad was right there feeding it. The little crow, Roscoe, imprinted on my dad and from then on considered us his family.

Robert and I served as crow siblings. On seeing us, Roscoe would hop across the room and set to wrestling with our hands, squealing and cawing in mock fury as we tousled him round and about.

Soon it was time for Roscoe to learn to fly. We raised a perch on one side of the den and hoisted Roscoe high. Then we ran to the opposite wall and called him. He'd joyously leap off his perch and hop right to us. His hops grew longer, and one day he flew. We clapped and cheered and gave him a nice snack of raw liver as a reward which he guzzled down with his signature liver-guzzling gurgle.

I was enamored with Roscoe. I loved all the animals that came through our home, some as pets, some as rescues, some as one-night stands, but Roscoe was something else. I was experiencing the intelligence, loyalty, craftiness, and soon, the beauty of the American Crow up close. When the time came to send Roscoe out into the neighborhood to live as a real crow, I was scared. Would he fly away like so many other rescued birds?

Roscoe flew right up into the pine trees, where he was greeted by several curious crows. I anxiously watched those other birds, hoping they'd accept him. Very soon the meet-and-greet was over. The neighborhood crows liked their new friend just fine. Daddy called Roscoe, who flew down from the trees and lit on his shoulder, ready to go back inside for dinner.

This was the 70s, though, and indoor pets were not to become fashionable for another decade. Roscoe had to take up residence outside along with our dog, Bowzer. Roscoe and Bowzer grew thick as thieves. When Robert and I rode our bikes to Zippy Mart, where we spent our weekly 25 cent allowances on candy, Bowzer and Roscoe followed along, lying down beside and perching upon our bicycles, respectively, while we were inside. They chased the family car together, and even teamed up once to catch a rabbit.

I spent long hours outside with Bowzer and Roscoe, my two favorite souls on the planet at that time. No offense to my human family, but there comes a time in every kid's life when our pets take the larger place in our heart.

Roscoe was wily and playful. Sometimes I'd discover one of his caches of shiny treasures. I'd roll back a log, and just as I'd reach for a twinkling gum wrapper, Roscoe would attack my hand furiously, loudly, fiercely. I'd laugh and roll the log back over his treasures. Roscoe, forgiving, would hop up onto my shoulder, rub his head against my neck and purr. Crows, unlike other birds who sing loudly to court mates, sing quietly with soft growly, rattly sounds. This nuzzling and cooing was Roscoe's way of wooing me. Consider me wooed. I'd reach up and scratch his neck, which was shockingly thin underneath that thick feather cowl.

Later that summer, we took our annual trip to Orange Beach. Not wanting to abandon Roscoe, we brought him along (unlike poor

Bowzer, who was left to fend for himself in the yard, with a neighbor pouring him a bowl of kibble every day).

Roscoe hung around with us at our cottage on the bay and got along fine with the local crows. We left him outside at night, and again, I feared he might join the other crows for good. Lo and behold, with the sun's first light, Roscoe was right outside the porch cawing with all his heart for his breakfast.

Roscoe, having imprinted on my dad, had no intention of leaving. This made it all the more terrible when Roscoe was stolen out of yard, which we would piece together years after his heart-breaking, sudden disappearance.

Every crow I've seen since then reminds me of that magical time when Roscoe was part of our family. In my heart, they are all an extension of Roscoe.

State Parks, My OG Stomping Grounds
May 11, 2023

A few weekends ago, I crashed my brother's 42nd High School Reunion. Among the other activities, which included a slap-in-the-face awakening to our advancing age with a visit to an ear-splittingly loud college bar, was a picnic lunch at the upper pavilion at Chewacla State Park.

To prove we aren't OLD old, we hiked down to the bottom of the waterfall. Oh, how sentimentally delightful it was to take that trek with Dusty, my first high school sweetheart, and Susan and Lizzie, two members of my OG Squad.

(Senior interpretation: OG stands for "Original Gangster," but is used to mean simply "original." I employ it here to emphasize that we were called the Squad long before "Squad" became a word embroidered on pillows, painted on wine glasses, emblazoned on t-shirts, etc...)

When we, this group of 59-year-oldish friends, reached the bottom of the steep, root-crossed, rock-strewn trail, Susan asked some youngsters hanging out down there to take our picture. They obliged, amused with us and the defunct selfie-stick Susan thought to bring along.

It's a cute picture. There we are, all smiles, standing at the bottom of that waterfall, lost in memories. I personally was recalling how Dusty and I used to sneak beer into the park and sit in his jeep, listening to James Taylor and drinking Budweiser like we were some kind of blazing comets in a world of mediocrity.

It felt good to go down there and see Chewacla Falls again, even if it hurt coming up.

I grew up swimming in Chewacla Lake, and in adulthood have enjoyed spending a night or two in the Chewacla Cabins. Whenever I enter the gates of the park, the timelessness of the arched stone bridges and walls, and of the worn, soft trails here at the foothills of the Appalachians enchants me.

Chewacla Park was built by the men who were provided jobs and livelihoods as members of the Civilian Conservation Corps (CCC), established by FDR's New Deal. Between September 1935 and March 1941, the CCC built the dam that now creates the stairstep, cascading falls; the 26-acre lake; the roads; the cabins; the arch bridges; the bathhouse; and more. During this time the park was turned over to the State of Alabama with the enactment of the Alabama State Park System in 1939.

Today, Alabama's 21 state parks encompass some 48,000 acres of land and water. These parks range from the southernmost Gulf State Park in Gulf Shores to the northernmost Joe Wheeler State Park in Rogersville, and from the easternmost Lakepoint State Park in Eufaula to the westernmost Bladon State Park in Bladon Springs. There are oh, so many state parks in between.

I have an annotated and creased map of Alabama's State Parks tucked into my travel notebook, with phone numbers highlighted in yellow. While working for the Alabama State Department of Education and traveling throughout the state like an intrepid, manic pilgrim, I discovered that state employees are granted significant discounts to state park lodgings when on business.

What a wonder! My noisily air-conditioned overnights in the Comfort Inn in Florence gave way to the glorious serenity of Joe Wheeler State Park and velvet-nosed deer at my doorway. Instead of resenting the buzz and bumble of Interstate 65 outside my window of the Best Western in Birmingham, I could rest in a cottage on a lake in Oak Mountain State Park, which meant canoeing in the afternoons with Lizzie for my daily constitutionals rather than walking laps in a black and sticky hotel parking lot. I was even able to book a staycation in the Chewacla cabins when working in Auburn.

The last leg of my travel before I retired was to Mobile, where I stayed in a newly constructed cabin in Meaher State Park in Spanish Fort. After work, rather than staring at myself propped up to a mirror at my "desk" in a hotel, I wandered along a boardwalk overlooking spider lilies and alligators, pondering my next moves, the saltwater of the Mobile Bay filling my soul while the stars overhead reminded me that life is a mystery, and not something to be manipulated while staring over a computer screen into a mirror in a room at a Holiday Inn.

If you are a state employee, a veteran, active military, or disabled, you might qualify for discounts at overnight stays at Alabama State Parks. I suggest calling rather than using the internet to secure reservations. My experience supports this OG method of communication in securing discounts as well as the hearing delightful stories from desk clerks.

Of course, there are senior perks as well. Comes a time when the lapping of lake waters and the chirring of crickets beats the din of the bar downtown. If you are lucky enough to have reached this point in your life, you too can enjoy our extraordinary state parks at a senior discount.

These days it feels like vacation and recreation are a rich man's private playground. Thankfully, our state parks offer respite and recreation for us all.

Cahaba Lilies and Love in Bloom
May 18, 2023

"I want to see the Cahaba Lilies," I said to our bonus-son Joseph a few weeks ago. Cahaba Lilies, one of Alabama's guarded treasures, bloom in the Cahaba, Tallapoosa, Chattahoochee, Coosa, and Black Warrior Rivers between Mother's Day and Father's Day. We circled a date on the calendar, May 13, one day before Mother's Day, but the day that was best due to work-imposed schedules by which we humans are so tightly constricted.

The night before our float, Joseph and his sweetheart Aidan made supper for us: mushroom stroganoff from foraged fungi topped with the tender tips of smilax cut fresh from the brambles around the house. As we dined, my phone began to ding with messages suggesting that the river conditions could be dangerous. Experienced kayakers were pulling out of their scheduled trips on the Tallapoosa, and I grew uneasy with the thought of a flow rate of 10,000 cubic feet per second, even though up until that moment I'd never given one bit of thought to "flow rate" and didn't know a 10,000 from a 10.

After some nervous grappling, I backed out. My stepmother Janie backed out. My husband Joe, who had to work and was not going to begin with, was relieved, remembering an earlier unfortunate family incident on a river which ended well but could have had a very different outcome.

Joseph listened to the arguments and got on his computer. With his engineering brain and biologist brain working in tandem, he calculated the amount of recent rainfall in the watershed (none), and the rate at which the water would flow if such and such amount was released from the such and such dam. Using all sorts of other wizardry, mathematics, and magic, he emphatically pronounced the trip safe.

"You should trust me. I understand rivers," he insisted.

Still, I didn't want to risk it. After all, we could go another day, right?

"But not with Aidan while the lilies are in bloom," Joseph said, insistent that we all go together.

With the light of early morning, all indications from gauges, maps, and metrics now supported Joseph's claim. It would indeed be a great day on the river. The trip was back on. But first, we all had to gather around Joseph and sing, "You were right, we were wrong, that's why we are singing this song!"

The Cahaba lilies were just beginning to bloom. Each ethereal blossom felt like a promise to cherish. Cahaba lilies are rare, with populations occurring only in Alabama, Georgia, and South Carolina. These spectacular lilies, known as shoal lilies outside of Alabama, are threatened by dams, sedimentation, and poaching. They are currently under consideration for protection under the Endangered Species Act. Aren't we just ridiculously lucky to see them at all?

And oh, how gloriously beautiful they are with their luminous white centers and five radiant spires topping sturdy, lush green stems. William Bartram, naturalist and America's first professional botanist, wrote about them, "There is nothing in vegetable nature more pleasing."

Growing out of shallow shoals in swift flowing water, the lilies hang on tightly with deeply embedded roots. When they go to seed, the stalks bend, dropping the seeds into the water. The fast-flowing waters then anchor the seeds into rocky crevices and cracks, setting up for the next generation. The wilder the waters, the hardier the foundation.

Mid-way down the river, we pulled up on some rocks to enjoy a clump of Cahaba Lilies and to swim a bit in the fragrant atmosphere they provided. Joseph stooped down into the flowers and retrieved something from his pocket. Presenting a hand-carved mother-of-pearl necklace to Aidan, he asked her to marry him for love, as they'd already legally married for health insurance. She accepted, and there amongst these precious lilies, Joseph the River Rat and Aidan the Firefighter declared each other to be the river and the fire in each other's hearts and souls.

Well knock me over with a Cahaba Lily! No wonder Joseph was so disheartened at the thought of canceling the trip!

Our perfect day on the river just became more perfect, if such a thing is possible. We journeyed onward, letting the current take us slowly in places and speedily in others. The sun rose higher, lighting up the mountain laurel on the rock bluffs. River turtles, dragonflies, Canada geese, Wood Ducks and their ducklings, and Tiger Swallowtails showed themselves, reminding us that we are the visitors here. Birdsong and the tumble and rush of water over rocks provided all the music we needed.

Cheers to flowers that bloom in turbulent waters. And cheers to love, sweet love.

Magical Summer Nights Are Here
May 25, 2023

Welcome back, lightning bugs! Every year, I make note of the arrival of our bioluminescent friends in my little green notebook. I saw my first about a month ago, a single blink in the trees. These nights, however, the yard is a wonderland of flashes and twinkles.

Before I write another word, I must address the linguistic issue. Is it "lightning bug" or "firefly"? Like the great Stuffing vs. Dressing Debate that occurs annually around Thanksgiving, the lightning bug/firefly controversy has folks drawing lines and preparing for battle. Most of those polled in Alabama say lightning bug, while populations out west say firefly. Maybe it's because we have more lightning here and they have more wildfires there.

Whatever we call them, they aren't flies, and although we commonly refer to any little creepy-crawly as a bug, to an entomologist, a true bug is an insect in the order Hemptera, disqualifying lightning bugs as bugs.

So what are they? Lightning bugs are beetles, belonging to the order Coleoptera along with lady bugs and June bugs, also not bugs. I've heard a dirty rumor that some people say "lady beetle" and "June beetle." I'll never join those ranks.

As a child, I thought lightning bugs were magic. Not only do they light up, they are slow moving, they hover right at eye-level, and they land right on your finger if you're patient. Their friendliness is based on their lack of fear of predators. Among other messages, their blinking warns potential hunters of a foul-tasting steroid in their blood. What feels like a greeting is a firefly's attempt to blink me away.

Fireflies belong to a family called Lampyridae, named from the Greek word lampein, meaning to shine. There are over 2000 species of Lampyridae, and true to the nature of mixed-up science, they don't all light up. Not only that, not all light-up insects belong to the family Lampyridae.

There's the glowing cockroach, for instance. This insect does not create its own light, rather, it relies on a colony of bioluminescent bacteria to do the work. Before you start shrieking, know that their glow creates a pretty cute pattern, like a teddy bear. Also, glowing cockroaches are probably extinct, with only one specimen ever discovered, living on a volcano in Ecuador which has since belched and gurgled in an active eruptive phase.

Another light-up friend of note is the larvae of *Orfelia fultoni*, a species of fungus gnat. These glow worms are nicknamed "dismalites" due to their presence in the Dismals Canyon, a natural wonderland located in Phil Campbell, Alabama,

The first time I went to the Dismals Canyon, I had zero expectations. I saw the name on a paper map and was intrigued. The only visitor there, I descended into a magical world of neon moss, towering sandstone boulders, pristine old-growth forest, grottos, weeping walls, and waterfalls. I might still be there if someone had not appeared to usher me out before nightfall.

Since then, I have learned more about the canyon. Knowing it houses the country's largest population of a wee glowing larvae that can only be seen in the dark of the canyon, I made a promise to return for a night tour, which I fulfilled last week.

In the company of my childhood companion Lizzie and her sweetheart Flip, we made the trek to Phil Campbell. After spending the day in the canyon, hearts and souls flooded with the beauty and splendor of all we saw, we returned after dark to see the famed dismalites.

Tiptoeing down crumbling and rickety steps into a canyon fed by a 15-foot waterfall in the dark of night with a whispering clutch of other humans was in itself an experience well worth the four-hour drive. The beams of our red flashlights perfectly illuminated our paths.

The occasional white light from cell phones, the ability to turn them off flummoxing their handlers, was blinding.

Our guide led us to a particular section surrounded by moss-covered rock walls and instructed us to turn off all lights. As our eyes slowly adjusted to the very blackness around us, the dismalites began to glow. Soon the canyon took on the feel of dark night under a starry sky, the tiny glowing larvae painting constellations around us.

If you get close to one you can almost make out its shape, but when you shine a light on it it will cease to glow and quickly scurry into a tiny crag in the rock. In darkness, they are best illuminated by your peripheral vision. Don't look too hard at these Schrödinger's larvae.

As we creep into summer, I hope you all enjoy evenings lit with twinkling fireflies and starry skies. To encourage your own light-up companions, weed less, avoid insecticides, and turn off the lights. If inclined, I highly encourage a trip to the Dismals Canyon. Bring along a childhood friend who believes in fairies for added magic.

Cuttlefish Romance and Reproduction
June 1, 2023

June is here. As well as marking the onset of the academic summer, June is recognized as Pride Month, a time when some celebrate the myriad manifestations of human sexuality. But what about the other critters? Where's the love? This month, I'll share some other species' practices and habits that don't fit our preconceived notions of reproductive biology. Clutch the pearls! We're going to talk about the birds and the bees.

Let's begin with cuttlefish.

Cuttlefish are cephalopods, along with octopuses, nautiluses, and squids. Characterized by multiple arms surrounding a head (cephalopod translates to "head foot"), a muscular mantle (except the nautilus, which is the only one to retain a hard shell), and large brains, cephalopods are among the most fascinating creatures on the planet, and by far the most intelligent of the invertebrates, as defined by humans.

While octopuses get a good deal of media and pop culture attention, I find cuttlefish equally praise-worthy. Like their octopus relatives, they can change colors and patterns almost instantly, and dazzle with an array of flashy hues. In fact, they have the most dynamic and diverse color options of all the cephalopods.

I grew interested in these animals after my first and only encounter with a real cuttlefish at an aquarium in O'ahu, Hawaii. That little cuttlefish absolutely interacted with us as we walked back and forth in front of its tank. It would lower its eight arms downward, peer inquisitively at us with its curious, bulging eyes, then propel itself away, its arms swooshing outward and its delicate fins, which run alongside its body, fluttering like silken wings. Over and over, it would come back to us, as if we were the attraction that day. It took a lot of wifely persuasion to convince Joe that we did not need a pet cuttlefish.

As with other species, courting brings out some fascinating behaviors in rival male cuttlefish. They flash brightly. They pulsate. They challenge other males, rushing and fighting each other. And sometimes, they deceive.

One clever deception, seen in about 39% of male cuttlefish, is the ability to display male color patterns on the side of its body facing a female while presenting the other side of its body as female. "I'm just a female hanging out with the gals," is the message to all but the object of his desire. This two-sided talent gives the gender-bender more time to seduce his crush as the rival males remain unthreatened by the perceived presence of another male.

The giant Australian cuttlefish goes even further. When wooing, male giant Australian cuttlefish flash neon blues and purples, brilliant greens, ruby reds, and golden patterns of dots and lines. Chromatic dazzling is accompanied by posturing; a male will flare his arms in an attempt to look bigger, and show off his hectocotylus, a tentacle dedicated to delivering sperm, in front of his face like a healthy handlebar mustache. The larger males intimidate and chase away their smaller competitors.

Rather than feeling sorry for themselves, the little guys resort to trickery. These cuttlefish, known as "sneaker males," go full-on drag. They pull in their arms and change colors to appear female, tucking their hectocotyli under their arms. In this way they are able to sneak right past the macho cuttlefish and mate with the females, who accept these sneaker males about 60% of the time, while the big boys battle it out with each other.

All that's well and good, but there's more. Refresh your coffee for this part.

The cuttlefish mating act is a head-to-head event. Males will "punch" packets of sperm right into the females' mouths. Females store the sperm in specialized sperm cavities for up to five months. After mating with several males, the female cuttlefish will selectively fertilize her eggs, then attach them to sheltered rocks or crevices where they are then left to their own devices to thrive or die.

How a female remembers which sperm packets belong to which male is more than my human brain can comprehend.

I can tell you right where I was when I learned that cephalopods select stored sperm for fertilizing their eggs, just as I can tell you where I was when Auburn beat Alabama in the famous "Punt Bama Punt" 1972 Iron Bowl, when I heard Elvis died, and when Joe first told me he loved me.

If you were a polygynous female cuttlefish, mating with multiple partners, in the one and only mating season of your life, whose sperm would you select to best ensure the survival of your species?

Research by dedicated teuthologists, scientists who study cephalopods, indicates that females prefer the sperm of the "sneaker males." Teuthologists hypothesize that crafty intelligence is a better marker for cephalopod success than flashiness and size. It's brains over brawn for the giant Australian cuttlefish.

If you'd like to witness the giant Australian cuttlefish reproductive extravaganza, visit Point Lowly, Whyalla, South Australia between May and July, when hundreds of thousands of these splendiferous creatures come a'courting.

The Eel Question
June 8, 2023

I learned some new words today as I read up on the mysteries of eel reproduction, our second foray into the wild, wild world of animal reproduction in celebration of Pride Month.

For instance, I learned that a ribbon eel is a sequential protandric hermaphrodite, meaning it is born a male and transitions to a female.

If such a word exists, then surely its opposite does as well. Indeed, there is a word for organisms which change from female to male: protogyny. These are such straightforward words! "Proto" is first, "andro" relates to males, "gyno" relates to females.

Well knock me over with a red feather plucked from the male half of a gynandromorphic cardinal, with one side of its body male and the other side female! Critters have been going around changing sexes under our noses all this time. Various frogs, humphead wrasses, sea bass, hawkfish, clownfish, and ribbon eels are among these shifty animals. And biologists have been using five-dollar words to describe these changes with no fanfare at all, no outrage, no town hall hysteria, no boycotting, nada!

Armed with dispassionate, descriptive words, let's discuss freshwater eels.

As far back as Aristotle, zoologists have pondered "the eel question," that is, how do freshwater eels reproduce? Unable to find any eel sex organs, Aristotle proposed eels reproduced without fertilization, an act of parthenogenesis. Even Pliny the Elder had something to say about eel reproduction, but I was so distracted by memories of passing clandestine notes during World History class in high school at the mention of Pliny the Elder that I can't tell you what.

Later, Sigmund Freud, searching for testes, dissected hundreds of eels and found none. Of course he did, I thought, carelessly generalizing all his work into one singular penis envy theory.

Eventually, in 1908, a scientist named Johannes Schmidt set out on a research vessel named *Thor* to find some answers to "the eel question." For 20 years he cruised the waters, collecting, mapping, and analyzing his findings.

Thanks to Schmidt and other eel enthusiasts, we now understand the eel life cycle.

Freshwater eels are born in the Sargasso Sea. Their tiny, leaf-shaped larvae drift about in the ocean for up to a year, then they then grow into what we call glass eels, some 5-7 cm long. These glass eels migrate to freshwater ponds and rivers while transforming into elvers, or smaller versions of adult eels. Finally, they morph into yellow-brown eels, which we call "eels." In this stage, which can last 50 years, they are sexless, hence the trouble finding those sex organs.

Near the end of their life cycles, eels return to the Sargasso Sea, sometimes thousands of miles away, to mate and spawn. Be amazed that this lowly creature is born in saltwater, lives in freshwater, and returns to salt water to breed and die. This makes them catadromous, and the opposite of salmon which are anadromous.

During their last journey, they don't eat. Their stomachs disintegrate and their eyes double in size, giving an eely image of the expression, "My eyes were bigger than my stomach."

It is only in this final life stage, occurring in the Sargasso Sea, that eels become male or female. Their sex is determined not by genes but by external conditions including water temperature and salinity, and possibly the population density of other male and female eels.

The actual act of mating is only surmised. Eels supposedly swim about in a frenzy, the females laying the eggs and the males covering them with a cloud of sperm. Eel researchers have never witnessed this act. In captivity, eel breeders must chemically induce eels to grow eggs or sperm, and artificially inseminate the eggs.

Then there are ribbon eels, a type of moray eel found throughout the Indo-Pacific region living in caves and rubble. Bright yellow with brilliant blue dorsal fins along their bodies, ribbon eels are inarguably gorgeous as they slither through the seas in a graceful serpentine fashion, despite reaching lengths of up to 10 feet.

As well as gorgeous, ribbon eels are described as sequential protandrous hermaphrodites, as mentioned earlier. All ribbon eels are born male, displaying a striking color pattern: bright blue with a brilliant yellow dorsal fin along their entire bodies. The males are fine in each other's company, even sharing their hidey-holes in rocks and crevices.

In their final stage of life, all ribbon eels transition into females, accompanied by a color change rendering them solid yellow. The adult female ribbon eel will mate with a male and lay eggs, then die within a month.

I encourage you all to watch a few YouTube videos of male ribbon eels swimming. It's as if they wave their own Pride Flag for an underwater world that isn't bothered by sex changes in the least.

Let the Rodents Do the Work
June 15, 2023

With Father's Day nearly here, this seems as good a time as any to celebrate one of my favorite animals, *Castor canadensis*, the North American Beaver. Male beavers are notably excellent dads. Along with their lifelong, monogamous mates, beaver dads are hands-on parents, acting as role models for as well as defenders of their young. As a result, beavers have one of the soundest family structures among mammals.

When I was a youngster, our dog Ziggy disrupted a family of beavers. With a head full of maternal urges having recently weaned a litter of pups, Ziggy stole two baby beavers from a beaver lodge and adopted them as her own. Ziggy made the local news with her interspecies care, nursing the baby beavers as if they were her puppies.

Soon the kits graduated to bottles. Rating a ten on the adorableness scale, a baby beaver drinking from a bottle is unspeakably cute. Beavers are adept with their front paws, and these little beavers clasped the bottles just as human infants would and gazed into our eyes as we fed them.

We passed one of the babies on to a friend and devoted our attention to the remaining kit who we named Beave. Beave grew quickly. To ensure his feet developed properly, we took him swimming frequently. My brother Robert and I would swim apart in the lake, and Beave would swim back and forth between us, taking a rest on our shoulders, nuzzling our necks between laps.

We eventually released Beave into the wild. My dad, the infamous snake-man Bob Mount, would visit him often, bringing honey buns from the Zippy Mart. These visits ceased one fateful afternoon when my dad showed up empty-handed and Beave attacked him.

A beaver's teeth are covered in iron-rich enamel, giving them their trademark orange color as well as the ability to chomp down a small tree in just minutes. My father was not willing to have a round with those teeth, no matter how adorable Beave was.

Notice that I attributed a sex to this beaver. In reality, the only way to tell a male beaver from a female is to express its anal glands. The color and scent of the discharge reveals the sex. As far as I know, we never scientifically determined the sex of our beaver. Imagine with me Instagram-worthy beaver gender reveal events.

Despite the near cult-like enthusiasm of our new Buc-ee's, some consider beavers a pest. Environmentalists and other scientists disagree.

Beavers are a keystone species, necessary to prevent the collapse of their ecosystem. The near extinction of beavers in our nation's history has left a scar. We may never experience the streams and ponds in the deserts our beavers once created.

Between the fur traders of the 1600s and the following colonization of North America for farms, our beaver population plummeted from 400 million to about 100,000. This brutal assault resulted in the loss of an estimated 195,000 - 260,00 acres of wetlands. While beavers are no longer considered endangered, the same cannot be said of our wetlands.

As ecosystem engineers, beavers alter their landscape by building dams, an innate behavior triggered by the sound of running water. These dams create ponds and swamps which provide both food and shelter for beavers. Within the ponds beavers build lodges and canals, creating intricate wetland environments which are home to wood ducks, turtles, fish, migratory waterfowl, and countless other critters.

These wetlands are capable of mitigating climate change by soaking the water table which in turn helps prevent drought, creating firebreaks in areas threatened by wildfires, and replenishing waters in areas of drastic snowmelt. The ponds also act as carbon sinks, trapping and holding onto excess carbon from the atmosphere.

Additionally, beaver dams improve the quality of the water downstream, filtering agricultural contaminants and other pollutants by up to 46%.

Armed with an appreciation and understanding of the beneficial effects of beavers on our environment, efforts have been made in the last hundred years to reintroduce beavers to the landscape. My favorite of these was the Idaho Fish and Game's relocation strategy in 1948. Rather than exterminate these large rodents who'd become pests in agricultural areas, they dropped them by parachute, having a stash left over from the recent World War, into a heavily forested area. Contained in crates designed to open on impact, all but one of the 76 beavers relocated by parachute survived. In flyover studies conducted the following year, scientists were pleased to see new wetlands and canals, evidence of the beavers' hard work.

As well as gaining recognition for their important contributions to our environment, beaver engineering has proven to be a cost effective, efficient method of restoring wetlands and managing flooding. "Let the rodent do the work," as those in beaver restoration say.

Happy Father's Day out there to all the dads who, like the talented beaver, get so many important jobs done!

Life Finds a Way
June 22, 2023

Ah, life! It just keeps rambling on, its celebrants mixing it up with all sorts of combinations and mechanisms, keeping us on our toes and sometimes, in our laboratories.

Lately I've seen news of a crocodilian virgin birth splashed across my headlines. Yet another example of the birds and the bees gone wild, I think, and set about to ponder the many methods of reproduction.

Back in the day we traded insults generously. One of our favorites was, "You were born in a test-tube!" Then in 1978, when I was 14, a baby was born who was conceived via in vitro fertilization (IVF) and that insult was put to rest. Despite the media-hyped fear mongering over "test tube babies," at the time, today about 55,000 children are born via IVF annually in the US. Other than the sticker-shocked parents, no one bats an eye over IVF.

Fast-forward to 1996. I was 32 years old with three daughters of my own. In July of that year, a sheep named Dolly was born in Edinburgh, Scotland to one parent. The first cloned mammal competed with the Summer Olympics in Atlanta for headlines.

"Oh, look, a cloned sheep," I said to my husband Joe. "Can you go pick up the pizza?"

We were other-wordly busy. Brand X Pizza on Thursday nights was supremely important.

Years later I returned to thoughts of Dolly. How the heck did they clone a sheep? It happened something like this. Scientists extracted the DNA from a mammary gland cell of a white-faced sheep and inserted it into an egg cell, which had had its nucleus removed, of a black-faced sheep. Recall from the cell model you made in 7th grade that the nucleus contains the DNA. Thus, the egg cell was devoid of genetic material.

Next, in a move hearkening to Dr. Frankenstein's methods, scientists joined the mammary gland cell with the modified egg cell using electricity, and voila. The black-faced sheep's egg cell now contained the white-faced sheep's DNA. That same electrical pulse prodded the egg to begin cell-division, typically stimulated by hormones.

The egg cell divided until it became an embryo which was then inserted into a surrogate ewe, and five months later, Dolly was born.

These days cloned animals include cattle, horses, dogs, mules, oxen, rabbits, rats, and a rhesus monkey as well as sheep. So far humans have remained uncloned as far as we know. It has all to do with spindle proteins, which I'll leave you to explore on your own.

You can even get your beloved pet cloned. A clone of your dog will cost about $50,000. Your cloned horse will set you back about $85,000, but your cat will cost you a mere $35,000.

While I might be tempted to clone my beloved pets if I was richer in dollars than sense, there are no guarantees. In 1999, a Texas couple had their Brahman bull, Chance, cloned. Chance was a lovable, face kissing, snuggly bull, so incredibly sweet that even Mother Theresa had her picture taken with him. The owners decided to have him cloned. While they waited for the birth of the clone, Chance died, and his owner quickly set about to skin him right in the pasture so he could be taxidermied.

Chance's clone, Second Chance, was not so sweet. He aggressively gored his owner twice, tossing him in the air, slamming him into a fence, even ripping the poor guy's scrotum. Still, his owners loved him for his DNA.

You can listen to this bizarre story on the podcast, or what we used to call radio show, "This American Life." Search "If by Chance We Meet Again." Enjoy with a strong cup of coffee.

This brings us to today's news, in which Coquita, a crocodile in Costa Rica, experienced a "virgin birth." While this item just appeared in my newsfeed, the actual event occurred in 2018 when Coquita laid a clutch of 14 eggs without having been near a male crocodile in over 16 years. Using a simple flashlight, scientists were able to determine that seven of the eggs were viable and artificially incubated them. One developed into a fully formed crocodile fetus, but alas, did not live.

I don't know why it took five years for the next step to happen, but a report published in *Biology Letters* on June 7, 2023, revealed that the stillborn crocodile was not a result of delayed fertilization, but of parthenogenesis, a process in which a female reproduces without a male.

Parthenogenesis occurs in two ways, the most common of which occurs when an egg cell merges with a polar body, a by-product of meiosis. Put simply, a viable egg cell merges with an inert, cast-off egg cell, instead of a sperm cell.

Parthenogenesis, common in wasps, snakes, sharks, and other species, is on the rise in the animal kingdom. We don't know why. As Jeff Goldblum's character says in Jurassic Park, "Life finds a way."

Happy first full day of our astronomical summer! Enjoy the sun, the original source of energy for all our crazy life processes.

Ziggy's Story
June 24, 2023

At any given moment on any given day, you might catch me thinking about dogs. Right now, I'm thinking about Ziggy, introduced here a few weeks ago as the mama dog who stole the two baby beavers. That alone is story-worthy, but it turns out Ziggy has her own story which deserves to be shared.

Her story is attached to my story, of course, so I have to start with me, age 12, disillusioned with life and angry at the world. My parents were recently divorced, and just as I was welling up with enormous feelings about that, the stepparents arrived on the scene. It was integral to my identity to hate them. They gave me a very convenient container into which I could pour my anger.

No matter how hard I tried to hate Janie and Steve, however, they both loved dogs, and this was something I could not compromise on. I love dogs. Period. This made my hating difficult, and at times, I shared warm moments with them over canines. Still, I was a sulky thing, in between dog episodes at least.

One day Janie showed up at "my dad's house," which had been the family home, now transformed, no longer the same place at all despite the same azalea bushes, the same club house, and the same dog Bowzer in the yard, with a big black and tan dog. I was intrigued.

"This is Ziggy," she said. I fell to my knees and hugged Ziggy. Ziggy gave me a tentative kiss. In that moment, I had no room for hatred, anger, disillusionment, or sadness. There was only this new dog. This Ziggy. This scrap of a life.

Ziggy was covered in the worst case of mange I've ever seen to this day. Her eyes were infected and white pus oozed out of all corners of each eye. Her body was emaciated, her ears tattered, her teats dangled like wadded up bubble gum.

"Hello, Ziggy," I said. I loved her immediately.

Ziggy, it turns out, had been missing for two years. Janie had been living in a duplex on Harper Avenue, a graduate student, when Ziggy was stolen from her fenced in yard.

Janie searched high and low for her dog. By the time she and my dad got married, she'd resigned herself to the fact that Ziggy was gone for good.

Then one day she got a call from a fellow teacher who'd seen a Doberman out at the pound. Jane was there in a flash. Sure enough, it was Ziggy. "The vet was supposed to come put this one down yesterday, but he never showed up," the attendant told her.

"How did you recognize her?" I asked Janie.

"She was my dog," Janie said. Anyone who has ever owned a dog understands.

In addition to her ragged appearance, x-rays confirmed that she was full of buckshot. No doubt Ziggy had been living off stolen chickens and pilfered trash for some time.

Ziggy's condition required constant care. I can still smell the milky-white wash we used to bathe her at least daily. We treated her oozing eyes with thick yellow ointment, washing the nasty "grahdoo," as we called it, off her face multiple times a day.

My brother Robert and I both took to caring for her without hesitation. Our dog Bowzer saw only her inner beauty and loved her on the spot. Slowly, Ziggy's coat transformed, her body filled out, and her eyes cleared up. Just to make sure Ziggy didn't get confused and wander off, we clipped her to a dog run when we left the house.

Then one day, my dad got a phone call in the middle of the night. "You've got my dog," a low, drunken-sounding voice said on the other end.

"No," my dad answered, "you stole my wife's dog when she lived on Harper Avenue."

Click.

Creepy as it was, that may have been the celebrated spark that incited my dad, a stalwart member of the Dogs Belong Outside Club, to bring the dogs inside. He was a quick convert, and to the end of his days, he not only allowed the dogs to be in the house, but also on the furniture, at the table, and of course, in the bed.

Ziggy's arrival heralded another shift: the end of the Only One Dog at a Time rule.

Now when we look back on the Ziggy story, Janie always remarks that watching how Robert and I accepted and loved Ziggy, warts and all, gave her a much-needed boost in confidence in her new role of step-mother, igniting a feeling that everything would be okay.

I'll go to my grave insisting that dogs are the most magical, mystical creatures on this planet, and that we, as a species, don't deserve their love. Ziggy's arrival in my life at a time when I what I needed more than anything in this world was to tend a soul more broken than mine only confirms what I already know.

Here's to dogs, who continue to work magic into my life daily.

Slow, Rare, and Indestructible
July 6, 2023

Alabama summer heat has arrived. Because the temperatures out there lately are extreme, and because I do love a connection, even if tenuous, I've decided to borrow some information from my online course, Keeping up with the Kingdoms: Extreme Animals I. Please enjoy these high-steppers in the extreme comfort of your air-conditioned home.

Before we begin, I'll share the categories with you in case you want to test your knowledge. Today we'll learn the world's slowest animal, the rarest animal, and the most indestructible. And yes, there will be a quiz at the end.

Let's begin with slowest animal, something we can all relate to right now as we sluggishly make our way across the kitchen for another glass of iced tea.

The winner of this category is the three-toed sloth. Surely a snail is slower than a sloth, you are likely thinking. Snails move at about 45 meters/hour, or 0.03 mph. Sloths travel at about 41 yards/hour, or 0.02 mph. Sloths win!

Why so slow, sloths? It's all about diet and digestion. Sloths primarily eat leaves, a low-calorie food providing scant little energy. In addition, sloths have large, four-chambered stomachs that take somewhere between 7-50 days to digest a leaf. With all that slow-moving food in their bellies, sloths experience a constant state of fullness. In short, they have very little energy at their disposal.

Sloths are so sedentary that they host their own ecosystems in their fur, including algae, fungi, and up to 150 sloth moths per sloth.

Next is the world's rarest animal, discounting those yet discovered and lone specimens in zoos. The vaquita, a small porpoise which lives in the north-west corner of the Gulf of California (Mexico), is currently our most endangered animal. What a terrible title to hold.

Vaquitas are recognizable by a black ring around their eyes and their proportionally large dorsal fins, necessary for dissipating heat from the warm waters in which they live. These small porpoises have seen a 90% decline in number since 2011 with an estimated population size of 8-10 as of this writing.

What could be killing off these shy little porpoises? Once again, it's human activity, specifically accidental bycatch by illegal gillnet fishing. To make matters worse, these insidious fisherfolk purposefully hunt another endangered animal, the totoaba, for their gill bladders, reputed to have great medicinal value and sold for top dollar on the black market. Sanctions by the Convention on International Trade in Endangered Species of Fauna and Flora (CITES) against Mexico due to their tolerance for this activity were recently dropped, prompted by promises by the Mexican government to enforce regulations to protect vaquitas, but CITES remains skeptical that Mexico's Environmental Department will keep those promises.

Moving along to a more cheerful topic, what was your guess for the world's most indestructible animal? The nuclear war surviving cockroach? The lethally venomous black mamba? Given their recent surge in popularity in popular culture, you might have correctly answered tardigrades, commonly called water bears or moss piglets. Tardigrades are tiny little invertebrates of the Tardigrada ("slow stepper") phylum, close relatives of arachnids. They look like tiny, fat bugs with eight claws and a pig-like snout and are less than 1.0 mm in length.

Tardigrades can survive the vacuum and radiation of space as well as scalding waters, freezing temperatures, and drought. When in deathly dry conditions, tardigrades go into a "tun" state of suspended metabolism which they can maintain for decades.

There are over 1,100 species of these wee animals, occupying every habitat on Earth, from tropical rain forests to the frozen

Antarctic to your yard. To see them, collect some moss, put it in a strainer lined with tissue, place the strainer over a bowl and add water so that the bottom of the moss is in the top of the water. 24 hours later, look for tiny wrigglers at the bottom of the bowl. Now collect them from the bottom by using a pipette, eyedropper, or pouring off the top layer of water. Examine your water droplets under a microscope.

If you don't have a microscope, don't panic. You can get a pocket microscope for around $15 which will do the trick and provide you with endless entertainment to boot. Search online if you can't find one locally.

As well as tardigrades, you're likely to spot nematodes, tiny, wriggling wormy critters. There are about 60 billion nematodes on Earth for every human, so chances are good you'll find them.

Each of these three animals is wildly more complex and interesting than I've described. In this day of easy access to information, I encourage all of you to learn more about them while waiting for the heat to pass.

The Toilet Snake Saga
July 13, 2023

World Snake Day is almost here! On July 16, please put aside a moment to celebrate our serpentine friends. To encourage you, I offer a tale of a very fine snake. Here's the story of the time we found a healthy midland water snake curled up in our toilet.

My husband Joe and I were out with the dogs when we got a text from our youngest, Anna.

"There's a snake in the toilet!"

She sent a photo. I was a little taken aback not only by its size but by how easily it seemed to nestle in the bowl, as if it found our potty comfortable. Not so our youngest child.

"Ugh! I feel so uncomfortable!" she texted.

Upon returning home and greeting our visitor, the first order of business was to identify the snake. I sent the picture to the experts on the Facebook group, "Alabama Reptile and Amphibian ID and Education" where she which was quickly identified as a non-venomous midland water snake. Next, I asked for removal advice.

Some of the suggestions were to dangle a fish on a stick, fashion a noose, grab her with salad tongs (which was my first thought, and which I attempted with near success), remove the toilet completely, and one of my favorites, fill the tub with water in hopes that she would simply relocate and we'd have a bathtub snake.

And then came the details.

- If you grab it, it will bite; it feels like Velcro being pressed into your skin.
- They have a weak blood thinner in their saliva, so the bite will bleed more than you'd expect.

- The bite won't hurt. 'Like a brier scratch,' your dad used to say. I'd be more concerned about getting musked.
- Water snakes will smear you with a nasty-smelling musk from glands at the anus. Not pleasant.
- Put on a light jacket, a parka will do, and some leather gloves and just grab it.

Armed with this information, Joe and I made several attempts to catch her, but with each try the snake retreated further into her hidey bowl-hole. I continued to check my Facebook thread in case anyone had The Magic Answer.

Folks were beginning to chime in to say they were moving not just out of our neighborhood but out of our town and possibly out of our state. Realtors take note.

I finally gave up and took off on a three-day girls' trip to Helen, GA to go tubing on the Chattahoochee. Snake in the toilet be dammed! Joe called a couple of friends and they continued the snake-removal mission into the night, finally giving up when the snake retreated and stayed gone for several hours.

Late that night, lying in my bed in the Helendorf River Inn, I checked the Facebook thread and found it had garnered almost 200 comments, with followers curious about what was going on with the snake. Our little snake, who I named "Hokey Pokey," was experiencing her 15 minutes of fame.

By morning Hokey Pokey had disappeared.

We worried over her health and safety for about two weeks. Just about the time we were giving up on her, she resurfaced, only this time she wasn't in the toilet. Joe found her hanging out in the rec room like she owned the place. At first he thought it was the fake snake I'd tossed onto the floor after deciding against hiding it in the bathroom to scare friends.

"I told the dogs to leave it alone," he told me, "but then the fake snake moved and disappeared behind a bookshelf."

For the next several days she showed herself briefly, but we could never catch her. Both Joe and Anna suggested releasing frogs and other small critters into the rec room in case she was hungry.

"She's like a pet now," Anna said.

We did leave a bowl of water out for her, but I drew the line at the frogs.

Finally, we saw her stretched out on a bookshelf, seemingly unaware of us. We quickly called Jennifer Lolley, a local naturalist, who came and snagged our snake with snake tongs and relocated her to the creek in our yard. I've never seen a happier snake. She took off into that water like a rocket.

That night Joe, Anna, and I admitted that we missed our toilet snake. The rec room felt incomplete with her quiet, mystical presence. Still, we were happy that she had been returned to her natural habitat where we continue to see midland water snakes regularly.

For those who want to know more about our local snakes, I highly recommend joining the Facebook group "Alabama Reptile and Amphibian ID & Education," moderated by the incomparable Raymond Corey who keeps the space cordial, interesting, and educational.

Mountains to the Gulf
July 20, 2023

I've just returned from the most remarkable trip!

I prowled in canyons in darkness and daylight, exploring cool limestone gorges and dark hidey-holes. I visited a working farm school with 40 acres of gardens and a handsome stud pig named Bo. I splashed in waterfalls, swam in hidden springs, hiked to a mountain overlook to bathe in the light of both sunset and sunrise, and wandered barefooted in crystal-clear rocky-bottomed streams. I bird-watched over a lake at dusk, searched for fluorescent lichens at night, and stood awestruck in a stand of longleaf pines. I dug my toes in the sands of the Gulf of Mexico and kayaked through the murky mud of a salt marsh, momentarily lost in the puzzle of man-made islands blanketed in nesting sea birds.

Did I mention, I waded through one of the world's largest white-topped pitcher plant bogs in the world?

During this week I was fed three ample meals a day and feted with wine and cheese at night. Not only that, but I was also chauffeured, free to read, nap, or even strum a guitar between destinations. I never once worried over arrangements of time or place, food or drink; it was all done for me, the drudgery and tedium of schedule changes around unexpected snafus being someone else's burden.

But wait, there's more. Traveling with me was a cadre of professionals whose vast understanding of paleontology, geology, herpetology, lichenology, hydrology, environmental policy, and all manner of -ologies was available, personally, at all times.

To add icing to the cake, we were met by other professionals at every stop, brilliant ambassadors thrilled to guide us through their favorite places.

Ah, this trip of a lifetime, this sampler platter of Earth's treasures, from mountains to streams to oceans! Surely, I traveled the world with lottery winnings, right?

In truth, I never left the state of Alabama, and it didn't cost me an arm or a leg. This glorious field trip, Mountains to the Gulf, is one of the outstanding teacher excursions offered by Legacy, an organization whose mission is "To be Alabama's primary source for science-based, environmental information and education, while cultivating a legacy of environmental stewardship for generations to come." As a member of Legacy, with the help of a scholarship from the Environmental Education Association of Alabama (EEAA), the total cost for my trip, including room and board, was $125.

I have neglected to mention that there were 19 educators, including but not limited to classroom teachers, on this week-long trip with me. Imagine the children throughout the state who will benefit from their teachers' exploration of Alabama.

Mountains to the Gulf had been on my bucket list for years. Back when I was a classroom science teacher, a week-long trip during my precious summertime, which was already filled with mandatory teacher workshops, away from my family was unthinkable. When I left the classroom for the Alabama State Department of Education, I was part of running those mandatory workshops, all summer long. There was no room for Mountains to the Gulf.

Time passed. The kids flew the nest. I retired and took a part-time job with Auburn City Schools, and suddenly the summer was mine! It didn't take a whole lot of arm-twisting for our bonus son Joseph to convince me to go along on this once in a lifetime trip, which he'd signed onto as both a participant and a herpetologist.

Now that we're all home, we stay in touch through our phones. Our Group Me conversations include full confessions that the trip changed lives, and suggestions that the same excursion be mandatory for all science teachers, and state politicians as well. Friends who followed our social media posts decry the exclusion of the general public. (Whispers amongst the Legacy crowd suggest that perhaps this may be a reality someday.)

While many shout-outs are due, Toni Bruner, the executive director of Legacy who was with us every step of the way, gets the first. Her tireless behind-the-scenes organization of Mountains to the Gulf made this experience possible.

It's going to take some time to process all that I learned about our uniquely situated state. Despite the notes I took in my little green notebook, I'm overwhelmed after "drinking from the firehose," as we said over and over, our brains swimming in contemplation of deep time and tectonic forces that formed our rivers, mountains, and valleys. It will be a while before I can package my new understanding of our state into digestible chunks. So for now, I'll revel in the sheer joy of being able to trek through the state with teachers and others who are truly the stewards of our rich biodiversity.

Searching for Elfland
July 27, 2023

When I was in high school, my friends and I spent a good bit of time in the Tuskegee National Forest at a spot we called Elfland. We didn't name this magical stretch of beach along Chewacla Creek, we just kept it alive, the name bequeathed to us from a generation of kids before.

Most weekends, you could find us there hanging out, swimming, playing softball on the grass farm on the other side of the creek. As night fell, we'd build a campfire and discuss philosophy deep into the wee hours. At some point those of us whose parents were keeping up the pretense of knowing where we were would pile into a car, drive along the bumpy dirt road to the highway, pull into The Torch Cafe and call our parents from the payphone outside.

I'd tell my folks I was spending the night with Dana. She'd tell her folks she was spending the night with Susan. Susan was spending the night with Lizzie, and Lizzie was spending the night with me. Somehow our parents bought it, or not and just let us be.

Oh, the conversations around the fire, stoked with Boone's Farm Strawberry Hill Wine and so many Virginia Slim Menthol Light cigarettes! We waxed on for hours under the stars before falling asleep, our sleeping bags right on the ground, dangerously close to the embers of the dying fire.

One morning I woke up, rubbed the sleep and sand out of my eyes, and spotted a canoe drifting down the creek.

"Look," I said, and poked my friends who stumbled groggily from their own sleeping bags. We watched in confusion turned amusement as two of our running buddies, good ole' boys who shall not be named, paddled by, their buck-naked bodies glinting in the morning sun. Why? This was the early 80s, and we were the last generation of feral kids. There is no other reason.

Tuskegee National Forest was our playground. We never stopped to think about the fact that it is a national forest, or what that means. It's only recently that I've paused to consider more closely what exactly Elfland was.

We have 155 national forests, 20 national grasslands, and one tall-grass prairie under the management of the U.S. Forest Service, together covering 193 million acres. The mission of the National Forest Service is "to sustain the health, diversity, and productivity of the Nation's forests and grasslands to meet the needs of present and future generations."

Our national forests fall in the "multiple use" category, providing space for hunting, fishing, swimming, camping, bike riding, horseback riding, birdwatching, and more within these working forests.

Four of our national forests exist in Alabama: Bankhead, Talladega, Tuskegee, and Conecuh. Tuskegee, my local national forest, is the smallest at 11,252 acres, some of the most abused and eroded wastelands in Alabama prior to their acquisition by the federal government in 1959.

I have friends today who take advantage of this forest regularly. It leads me to wonder, why did I quit playing at Elfland? When was my last time to wander down that craggly trail to emerge onto our pristine beach, covered in OFF! and sunscreen, eager to spend the day with friends?

This morning, I grabbed a water bottle, the obligatory bug spray, and a map of the forest and embarked on a quest to rediscover Elfland. Cranking Lynard Skynard tunes helped me navigate by intuition to the right road, where I turned confidently after a lapse of over 40 years. Checking the map for the spot where the grass farm and Chewacla Creek met, I tried to find our old parking area and trailhead. Neither the music nor my guts were of any assistance. A forest shifts and changes with age.

Could this be it, I wondered, stopping at the only space with any parking on the road. I searched for the tree where the boys carved their names, pulling vines back from tree trunks, finding no indication of their presence. I followed a short trail to the creek, but there was no sandy beach, no opening to the grass farm on the other side. I went back to my car and checked my map again, driving to the end of the dirt road and back, finding only that same parking place I'd already explored.

I parked again and wandered, trying to feel a worn-away path. Bright yellow mushrooms, crooked tree limbs, and a thousand spider webs refused to give up their secrets. If I want to find Elfland, I'll have to come back with a kayak and follow the creek. Maybe those good ole' boys know the way. Or maybe Elfland no longer exists. "Tuesday's gone with the wind," I sang to myself.

Despite not finding that sweet spot of magic from my youth, I enjoyed searching for Elfland, traipsing about the forest, my forest, your forest, our forest, that itty bitty sliver of our precious public land.

A Win for Our State Amphibian
July 3, 2023

Who doesn't love a salamander? Every kid I know is quick to turn over a log in the woods in search of these slimy, bulgy-eyed critters. Adults like them too. We hold them up to our faces and grin. They smile back at us with their big, friendly faces, eliciting all kinds of silly anthropomorphizing.

I'm pleased to say that we've got a lot of salamander species here in Alabama. We're home to spotted salamanders, striped salamanders, huge salamanders, tiny salamanders, salamanders with silly-frilly gills, salamanders with short legs, salamanders with two legs, and salamanders with no legs at all! Each of these little buddies deserves their own column, but today, due to events in the news, we're focusing on our state amphibian, the Red Hills Salamander.

This shy salamander lives on north-facing slopes of mature, closed-canopied hardwood forests down in south Alabama, in portions of Conecuh, Covington, Crenshaw, Butler, Monroe, and Wilcox counties. Red Hills salamanders, who spend their days tucked away in burrows, eluded curious children and scientists until 1960 when the first specimen was found. It was three long years before another was documented.

In 1977, the Red Hills salamander was listed as a federally threatened species, its survival impacted largely by, you know this one, habitat destruction.

The Red Hills salamander is endemic to Alabama, meaning that it lives here and only here. Cut off from other salamander species due to tectonic forces that shaped the land we now call Alabama millions of years ago, they are the only members of their genus. They are long, up to 11 inches, with very short legs, and have a deep translucent purple skin, the better to absorb oxygen through, my dear.

Red Hills Salamanders are one of the largest lungless salamanders in the world. Lungless salamanders? Yes, and raise a glass to them. Our terrestrial amphibians are indicator species, meaning they are constant sensors of the state of our planet, taking in their oxygen, and all the other stuff in the air, directly through their skin. Unfortunately, salamanders worldwide are in decline, a trustworthy measure of the state of our planet indeed.

Wait. Terrestrial amphibians? There are so many strange word combinations when it comes to Red Hills Salamanders. Think back to fourth grade, when you learned that amphibians are born in water, move to land, and return to water to lay eggs. Red Hills salamanders break the rules. They lay their eggs, a mere 6-10 per clutch, directly in their burrows, the eggs themselves full of the only watery home necessary for the development of the larval salamanders.

Here's another Red Hills Salamander fun fact. They are Alabama's only endemic terrestrial vertebrate. As I write this, another endemic salamander comes to mind, the Black Warrior Waterdog, yet it lives its entire life in water, making it an aquatic vertebrate. Salamanders are full of surprises.

Red Hills salamanders are decidedly cool, but the cool factor is not the reason Red Hills salamanders are in the news. In mid-July, a whopping 23,000 acres of forestland was acquired by The Conservation Fund, an organization dedicated to land preservation. This land, now called the Red Hills Salamander Forest, will remain intact and preserved as a healthy habitat for our state amphibian.

In addition to ensuring the best chances for the survival of the Red Hills Salamander, the Conservation Fund will allow hunting, hiking, fishing, bird watching, and other public activities in the Red Hills Salamander Forest. It's a win for the planet as well. According to The Conservation Fund, that 23,000-acre lot stores roughly 3.5 million tons of carbon dioxide, comparable to removing 779,000 gasoline-powered vehicles from the road for one year.

Another protected area, the Red Hills Tract, a 4,300-acre parcel of land owned and managed by the Forever Wild Land Trust, also includes Red Hills Salamander habitat. Though crisscrossed by private properties, this too is public land and accessible for all to enjoy while maintaining the integrity of the unique ecosystem.

My phone dings daily with calls and texts from land-grabbers wanting to gobble up my little undeveloped acre in town. None of these prospective buyers are interested in salamanders. Their eyes glaze over with the thought of wringing money out of clearing the lot and putting up ticky-tacky houses or apartments where the critters now live. This particular trend of land acquisition in our state is deeply disturbing to me. If you too are distressed by the willy-nilly destruction of our state's land, I suggest a deeper dive into the works of organizations like Forever Wild and The Conservation Fund.

Stay cool, drink your water, and enjoy the buzz of the cicadas!

Sturgeon Moon Float
August 10, 2023

I've always liked moongazing. When we were little, we used to lie in the yard with the dog and stare at the moon. "You'll go crazy," our parents told us, "turn into lunatics." We scoffed, but later, as a middle school teacher, I began to wonder about the moon's influence on behavior. Ask any teacher.

This August 1st, I experienced the full moon rising in the most glorious of ways, that is, on a nighttime kayak trip on a river on my favorite island with three of my very best long-term friends. The four of us, the OG Squad as mentioned in a previous column, gathered on Edisto Island, SC, last week for a week-long celebration of a 45 year-long friendship.

We'd been talking about this trip for a year over WhatsApp, which we use to communicate on an almost daily basis. Imagine the thrill when we went to rent bicycles for the week and happened to be right there when Carolyn at Edisto Water Sports and Tackle decided it'd be a swell idea to host a full moon float. We were in like Flynn.

This moon, the Full Sturgeon Moon so named by the Native Americans for the abundance of sturgeon in the Great Lakes at this time of year, was also a Super Moon. Super Moons occur when the full moon is at its closest point in its orbit around Earth, the perigee if you need a word for a trivia game. To the naked eye, the very slight differences in brightness and size are negligible, but the tides feel the pull and are more extreme at these times. All that to say, lots of water in the river and a big fat moon rising was a good excuse for a nighttime float.

We gathered with other lucky ducks who got a spot on this inaugural full moon float and one by one, we plunked down into our kayaks and pushed out into Big Bay Creek. Tiny lights on our paddles blinked and twinkled like crazed alligator eyes as we cruised up the creek with the incoming tide.

David, our guide, stopped us at a 2,000-year-old Indian Mound, a midden filled mostly with oyster shells but also some bones and pottery. A midden is a fancy word for a trash heap, another good word for your word bank. The artifacts discovered in this mound tell tales of a peaceful, nomadic tribe of hunter-gathers. This mound is being reclaimed by Mother Nature, and we are lucky to be here at this time while we can see it. Fascinating, yes, but this is not where my mind was.

I scanned the horizon. Where is the moon, I wondered? Certainly, it was not hiding behind some garish high-rise, as those blights are not allowed on Edisto Island. The night grew dark. We paddled up Scott Creek, through channels that were filled to the brim, connecting with other channels and sloshing the marsh grass underneath with the unusually high tide created by this Super Moon, the same moon that I could not find in the sky.

"Where is the moon?" I asked my friends Dana, Lizzie, and Susan. They too were searching. Others in our group began asking the same question. And then she appeared, a bright red dot peeking over the southeastern horizon. All eyes stared, many fingers pointed, and as she crept higher, revealing her perfectly round symmetry, bold and orange and growing huge, we broke into giddy applause.

Lunatics all, clapping and cheering the rising Sturgeon moon from our watery vantage point! She responded, shone brilliantly, casting orange reflections on the water like a glorious sunset. Our kayaks, freed from our attention, drifted into the marsh grasses, disturbing little birds. The fecund odor of pluff mud arose as we pushed ourselves away from the oozy, muddy marshes.

And then it was time to go. We'd come to see that full moon rise, but now the tide was literally turning, ready to assist us on our long way back to the pier.

In the darkness we drifted apart and enjoyed a spooky quiet paddling in the night, all alone, the quiet dipping of the oars breaking the line between dark water and sky. Full disclosure, not everyone in our party liked being adrift in the night, seemingly lost under a full moon in a dark river winding to the sea. But oh, I did.

One of my daughters says I might consider eliminating the word "should" from my conversational vocabulary. Still, I will employ it here. If you ever have the chance to see the moon rise from a kayak on a black river with your very best friends on a dark, summer night on a magical island, you should. You absolutely should.

Sipping Moonshine in Kenya
August 17, 2023

When I was 21 years old, I spent a month in Kenya as a volunteer research assistant to a professor who was studying the habits of the few remaining black rhinoceroses in Maasai Mara, data the Kenyan government needed to build a sanctuary for these endangered beasts. I have many stories to tell from that summer, but seeing as today is National Bootlegger's Day, I'm going to reminisce on a story about moonshine.

Granted, Bootlegger's Day marks the birthday of Al Capone as well as the first day of prohibition and does not extend to events on the other side of the globe, but I'll take any connection for a good story, tenuous or tight.

We set up camp on a gentle hill near a tree in Maasai Mara, 19 of us in all including 4 ascaris, or guards; a cook; the professor and his son; and us volunteers. Shadrack, our cook, worked wonders with posho and Sukuma wiki, fancy words for corn meal mash and braised collard greens, but still, we accented our meals with an occasional goat which we would purchase from the local Maasai, nomadic herding people who live in East Africa.

Buying a goat was one of my favorite activities of the summer. We'd pile into a Range Rovers, sides dented and gored by animals, two spare tires and a can of gas loaded and ready, to search for a goatherd.

One afternoon, we found an exceptionally tall young man, clad in a yellow robe and holding a long, crooked staff, with his goats. After much lively discussion in Swahili between him and our professor, nicknamed Doktari, and an exchange of a Polaroid picture, we were directed to the young man's home to make a deal with an elder.

The Maasai live in communities called Manyattas, consisting of rings of dung huts encircled by thick, thorny brush. This arrangement protects them and their herds of cows or goats from hungry lions at night. Most of the time, when dealing with the Maasai, we stayed near the vehicle while Doktari went inside the Manyatta for a conversation which could take five minutes or five hours.

This time we were invited into the Manyatta, and incredibly, into one of the huts, an unexpected, and according to Doktari, unprecedented move. We had to duck to enter the tiny home. Once inside, our eyes stung with smoke and we were overwhelmed by the sweet aroma of honey and milk. The largest room in this three-room structure was the kitchen, where an ancient woman crouched beside a small fire, stirring a pot, surrounded by family members. Two bedrooms, one for the husband and the other for his wives and children, and I suppose, the elderly woman in the kitchen, squared off hut.

We were directed to sit in the circle around the fire with the family. My Swahili was too rudimentary to understand what Doktari and the man of the house said. There was laughter, merriment, and smiles all around, but not from the old woman. She just glared.

Then out came the moonshine, straight from the urine-cured gourd with the beaded leather strap. We were each given small glasses of this frightening concoction. I took one sip and my eyeballs burned, my throat constricted, I thought I would cry. No sipping moonshine in Kenya for me, I decided. The others drank their moonshine, and I knew this would be one of those five-hour conversations.

Quietly, I slipped out of the hut, into the circle of the manyatta. The orange sun was setting and I heard cowbells in the distance. The clanging grew louder, and in came the goats, followed by dogs, and the tall yellow-clad man we'd met earlier. I stayed there until the sun set, a stranger in a strange land indeed, my mind boggled by the unknown complexities of other people's lives.

Eventually, Doktari and the others spilled out of hut, drunker than Cooter Brown. Squeezed in with the goat and three rowdy mates, I wondered if we'd make it back to camp at all.

"Is the goat speckled?" came a voice from camp, crackling over our radio, sending my drunken companions into uproarious laughter and our vehicle into a small spin.

Doktari would've been arrested for drunk driving if there'd been a road, or a law enforcement officer, or a law.

When we got back to camp I headed straight for my tent, grabbed my journal, and wrote a stream of thoughts about this magical, moonshiney night. And here I am, 48 years later, writing about it again.

Happy Bootlegger's Day to all who celebrate! May your moonshine send you only into tailspins of happiness!

Freshwater Mussels' Surprising Feats
August 24, 2023

When I was little, some of my favorite days were those we spent at Mussel Creek, our nickname due to the abundance of freshwater mussels found in its water, for the Uphapee Creek. Despite their presence, I never thought much about all those mussels.

Then one day last year, our bonus son Joseph began rambling on about mussel lures. I thought he was talking about fishing, so I responded with some inattentive nods and "uh huhs."

"No," he finally said, "not fishing lures. Mussel lures. Their bodies have lures shaped like little fish."

I'm well familiar with alligator snapping turtles and their wiggle-worm tongue lures, and deep-sea lanternfish with their alluring bioluminescent lures, but mussels? They don't even eat fish. Why would they need lures? Time for a fresh cup of coffee. It's going to get weird.

After fertilization, mussel eggs develop into larvae which are released into the water, specifically the water that flows through the gills of fish. If that's not weird enough, mussels have evolved a way of tricking the fish into gulping the larvae straight into their mouths by packing the babies into lures. These lures, extensions of the mussel's bodies, are brilliantly designed to mimic various tasty fish snacks.

Y'all. We're talking about mussels. With bodily extensions that look like other animals. How does that happen?

Imagine a moment, 245 million years ago, when some random mutation in some random freshwater mussel caused its mantle, the tissue between its body and the shell, to build a little protein flap that extended beyond the shell. When a fish came to eat that little flap, enticed by what it thought was a minnow, it instead got a mouthful of mussel larvae. Those larvae latched onto the fish's gills where they were well fed by the nutrients in the water passing through. The fish, in reaction to the presence of foreign bodies in its gills, formed cysts around the larvae, thus protecting the wee baby mussels. After some a few weeks, those mussels fell from the fish onto new territory, upstream from their mama mussel, where they grew into nice, healthy mussels with a weird mutant gene responsible for a bit of flapping tissue on the mantle.

These mutant mussels, with the aid of the fish, survived better than their peers and passed along that weird mutation to new generations. Millions upon millions of mutations later, some causing changes that made the flap appear more prey-like, such as a dog of pigment that resembles an eye or a streak of pigment that resembles a stripe, we have the current state of mussel lures. As well as displaying impressive mimicry, different mussel species have lures specific to the fish that live amongst them.

Oh my heavens, these lures, extensions of the mussel itself, are something else. Some look like darters, with stripes, eyes, fins, and tails. Some look like crawdads, with pointy faces and sharp legs. Some appear detached from the body of the mussel, little packets of larvae that strongly resemble minnows, strung along with a small bit of mucus from the maternal mussel. All these lures are loaded with larvae, waiting for their fish-vehicle.

Impressive lures aside, it's the work of the freshwater mussels in keeping our streams and rivers clean that makes them fundamentally valuable to the health of our waters. Mussels funnel in water for nutrients, removing organic particles of algae, bacteria, and other detritus, leaving clean water as a byproduct. A single adult filters roughly 15 gallons of water a day. One healthy mussel population in the upper Mississippi River was found to clean more than 14 billion gallons of water a day.

Unfortunately, freshwater mussels are facing a perilous decline. Largely due to the construction of dams, with other factors causing habitat degradation factoring in, freshwater mussels are among the most imperiled living things on the planet.

Alabama, our incredibly rich biodiverse hotspot of a state, is home to the greatest number of mussel species in North America, with 58 of our 181 native mussel species endangered or threatened. My childhood Mussel Creek is critical habitat for Finelined Pocketbook, Southern Clubshell, and Ovate Clubshell mussels, and historic habitat for Alabama Moccansinshells.

How heartbreaking to learn about incredible critters only to read that they are rocketing out of existence. Fortunately, the bipartisan "Recovering America's Wildlife Act," or RAWA, popular among Americans of all stripes and colors, offers protection for all threatened and endangered species, and restoration of species of greatest conservation needs. Including mussels.

To stay abreast of this bill, you can sign up for alerts by creating an account on Congress.gov. Search for S.1149 – Recovering America's Wildlife Act of 2023, click on it, and click "Get Alerts" under the title.

It'd be a shame to see such elaborate and beneficial critters, so many residing here in Alabama, disappear before our grandchildren can enjoy them.

Sharing the Love of Slime Molds
Aug. 31, 2023

Several weeks ago, I was out walking in the yard wondering where the spiders were when I spotted a bright yellow blob on a large, flat tree trunk. My heart skipped a beat and I rushed to inspect it. Sure enough, it was a resplendent specimen of a healthy dog-vomit slime mold, with nice clumps of protoplasm giving way to fan-like tendrils on one side.

"Hello you gorgeous thing!" I said out loud, wishing my loved ones were with me to share my excitement. They'd understand. When I first learned about slime molds I forced my family to watch YouTube videos and celebrate the wonders of this very strange organism with me. My online students in my "Keeping Up with the Kingdoms" class were appropriately awed as well.

You, fellow readers, have somehow been spared my slime mold fancy up to now. However, the time has come.

Let's begin.

If you've spent any time in the woods, wandering among damp and decomposing logs, you've likely encountered a few of the over 900 species of slime molds. Some appear as bright red BB-sized spheres clustering on tree stumps. Some are orange and stringy, like a popped bubble-gum bubble, and some look like bright yellow clumps of scrambled eggs or dog vomit. Likely, you mistook these amorphous blobs as alien fungi from another planet.

If you slowed down to study them, maybe even taking out your phone and pulling up your iNaturalist app to identify them, then followed up with a quick Google search about slime molds, you might have dug deeper into your belief that they are aliens.

So, what are they? Despite the misleading name, slime molds are not molds at all. They are not even in the fungus kingdom. Slime molds are protists, the lovable oddballs of the living world which include algae, amoebas, diatoms, and other organisms that find no home in any other kingdom.

Slime molds spend much of their lives as single-celled organisms, just hanging out living life. But when they find a good source of food, things get freakier than a Cici's Pizza buffet with chocolate pizza. Some slime molds will begin to grow and divide through a process of cell division called mitosis, which you learned about in middle school. However, they don't undergo cytokinesis, that final step of splitting into two cells. Instead, they end up fusing into a macroscopic, oozing, giant of a cell with hundreds of nuclei.

Well knock me over with a Biology 101 syllabus. I had no idea that one cell could have hundreds of nuclei.

Now this oozing slime mold will grow and stretch and pulse as it slimes its way around a fallen log, or other host of food. If you are lucky enough to find one in your yard or woods, you might want to check on it daily, watching it shift its shape and position, grow and shrink.

And then one day it will be dried up, with little dots left behind. Those are the new slime molds, all reproduced and ready to begin again.

It's impressive enough, this oozing, multinucleate mass, but we've only just begun.

These brainless blobs exhibit behavior suggesting the complex processes of planning and learning. In labs, researchers situated large petri dishes over maps of cities like Tokyo, and placed bits of oats on the points of interest. Slime molds, set loose in the dishes, detected the food and sent out tendrils to scout the most efficient pathways from one bit of oats to the next, rearranging, extending, withdrawing cytoplasm until the slime mold resembled a map at least as efficient than the public transportation systems laid out by civil engineers, and in some cases, superior.

But wait, there's more.

Slime molds are repelled by salt, though it's harmless to them. To test their perseverance, researchers placed salt barriers between slime molds and oats. Eventually, the slime molds couldn't resist the oats and they crossed the salt bridge. Realizing no harm was done, they boldly oozed through the salt in subsequent trials. Did they learn?

Learning would be a feat for a brainless plasmodium. Passing down learned behavior would also be incredible. And yet, successive generations of these intrepid traveling slime molds also crossed the salt barriers without hesitation.

I am not alone in my appreciation for slime molds. A paper published in *The Astrophysical Journal Letters* (March 2020) described the work of a team of scientists who are using the growth patterns of slime molds to model the "cosmic web," which makes up our entire cosmos with interconnected filaments of dark matter.

As for the fabulous specimen in our yard, it began to dry out quickly. After two days, I found a tiny snail munching away on its remains, which pleased me so much that I made a video of it and put it on Tik Tok.

As for the missing spiders, I'm perplexed, and wondering if my orb weavers of last year were just early.

Return of the Child Within
September 7, 2023

I just overheard the strangest confession coming out of my mouth.

"I don't know why," I said, "but all I want to do is play. Go kayaking. Ride horses. Read. I don't want to clean my house or work in my yard." I was speaking to my doctor on the state of my mental health.

He laughed, but my words took me aback. It seems I have reverted to being a child.

Not just any child, this is a child I recognize. Those things were my childhood dreams. It got me to thinking, when did I stop doing all those things, and why am I now so drawn to my childhood passions?

I remembered something I read years ago in a book, *The Madwoman in the Volvo: My Year of Raging Hormones*, by Sandra Tsing Loh. First, I recalled physically reading the book, crouched down on the backseat floorboard of the car, the only place I can read without getting carsick, on the way to New Orleans with my husband Joe and our friend Peter. Despite the excitement of the long weekend ahead and the scintillating good company, I couldn't tear myself away from this well-told tale of what was happening in my body.

The bit that was tingling my neurons into a memory now was a suggestion that women, affected by hormones, lose their identities during their fertile years, but Glory be, return to their true selves once they get all that behind them.

There is a world of articles out there on the power of older women and the anthropological importance of grandmothers in our development as a species. I've read a lot on the subject, as I'm sure most women my age have, and none of it surprises me. I've grown up surrounded by unconventional women who grow stronger and more resplendent with age. But this bit about the return of the childhood self is something new to celebrate.

Who would my childhood self be, given the space and time? It turns out I wouldn't be some famous author. I wouldn't have a well-tended natural yard of native plants. There'd be no whimsical porch or tidy house, and certainly no glamour or glory of any kind. I know this because for the first time in my adult life, I just had a real summer vacation, and I got to experience the return of the child within.

The most exciting development has been reuniting with horses. Always a "horse girl," I gave up riding when I moved to Massachusetts, and what with motherhood and full-time work, never picked it back up. Last spring, I began grieving my first horse, who died just four months after I got him, to anyone who would listen. My 13-year-old self was devastated then, and she was moving into my head now.

While at a high-school reunion, I began weeping in my beer to Lisa, an acquaintance, now friend, about missing horses. Lisa responded, "You should come ride our horses," and gave me her number. With that my life shifted, and now, there are horses upon horses, evenings of loping through pastures chasing cows, mornings of quiet trail rides and squeaky leather saddles.

I also bought my first kayak this summer, long overdue for someone who loved being in a river as a child. With every column and magazine article I wrote, I justified the expense. "Clickety-clack, kayak, clickety-clack, kayak," the keyboard said to me. Paddling through wildflowers, into sunsets, in rain and sun splashed rivers, with old friends, new friends, and total strangers has been a balm to my soul.

What is a return to childhood without children? We kept the granddaughters a good plenty. I dragged all my Breyer horses out of storage, and took the girls horseback riding. We toured the Auburn University Equestrian Center; they got their first cowgirl boots.

And there's more! I took a week-long bus trip through the state with a bunch of raggedy nature lovers during which we changed in and out of "wet clothes" so often I was able to revert back to being able to put on a wet bathing suit happily. At home, family gathered, and we hosted game nights and wine tastings with friends. I took a weeklong trip to Edisto Island with girlfriends and met up with more girlfriends for dinner once a week. I read a bunch of books and still got to hang out with Joe and the dogs on the couch most nights for a movie.

September is here, and my whirlwind summer vacation is over. My house is not clean, my yard is far from kept, and I'm still tending to my younger self, riding horses and kayaking. As for my keyboard, the sound is now clippety-clop, clippety-clop, keys clicking to subsidize horseback riding lessons for our granddaughters.

I'm curious and intrigued by women of a certain age returning to our childhood selves. Who were we then? Who are we now?

Relating to Flamingos
September 14, 2023

One of my student workers complained recently that her professor had made all members of her class tell everyone what animal they related to that day. I thought of the ice-breakers and warm-ups in the myriad professional learning sessions I've attended and conducted, and shook my head in sympathy.

"That's never gonna stop," I said, and then, because it's in my nature to yank a chain now and then, I had all the workers tell what kind of animal they related to that day.

We were a regular menagerie, including a fox, a goldfish, a cat, a pufferfish, and a bone-tired flamingo. "I'm on my last leg," she sighed, exhausted with exams.

We talked about our visiting American flamingos over in Hale County, blown off course by Hurricane Idalia. Alabama's not alone in welcoming these transplants; tempest-tossed flamingos have landed in 10 states, as far north as Ohio.

"Where do flamingos even live?" one of the workers asked.

I answered confidently, "They live in the tropics," but as the words were leaving my mouth, I realized I didn't know that for sure, just like I didn't know why they stand on one leg. It seems the only thing I actually knew about flamingos is that their color is derived from their diet, and they are noisy.

So off I went to find out more about these spectacularly beautiful, if odd, birds. I was right about their native habitat. American flamingos, one of six flamingo species, are permanent residents on the Caribbean islands, the Yucatan peninsula, and the northern coasts of

South America. All flamingos live in briny waters, lagoons or lakes, in habitats not suited for other birds.

A once-robust population of flamingos in Florida was decimated in the early 1900s due to habitat destruction and over-collecting of feathers and eggs.

I suppose a flamingo egg would be as good an egg as any. They are still sold as delicacies in some places. In ancient times, upper-crust Romans ate flamingo tongues at opulent banquets and feasts. For what it's worth, flamingo tongues contain erectile tissue, used for stabilizing the tongue when flamingos eat, which is an upside down, sucking in, pumping out affair leaving tiny crustaceans, algae, larvae, and other micro-bites trapped in the beak.

I'm a little worried about our visiting flamingos. Not about their food, I think they'll find wee critters aplenty in their lake, but about their socializing. Flamingos are colonial obligate birds, meaning they stick together all the time. A group of flamingos, when showing off for mates, is appropriately called a flamboyance. At other times, it's just a stand. These stands can contain over 10,000 birds.

Flamingos mate for life, which can be up to 20 years. The female will lay one egg on a mud nest, and both parents take turns incubating it. As I said, flamingos are noisy, and the chicks, while still inside the eggs, make distinctive calls that their parents learn to recognize. Once hatched, both parents care for the chick for about 75 days.

And now let's talk about lactation. If you'd told me that the mother birds of some species feed their chicks milk of any kind, I'd have said you didn't pay attention in biology class. If you'd gone on to tell me that the fathers did as well, I'd have said you skipped a whole unit all together.

Well knock me over with a pink flamingo feather. It turns out that indeed, parents of some birds begin lactating, sort of, producing milk in their crops just before their eggs hatch. This life-sustaining liquid, called crop milk, is a protein and fat-rich fluid excreted from the lining of the parent bird's digestive tracts. The parent birds regurgitate this nutrient-filled elixir and slide it into their offspring's beak. A video of this process went viral on the internet, with commenters horrified that a male flamingo had pecked the mother's head, and blood was flowing over the head and into the chick's mouth. In fact, this was just dear old dad, giving his baby some nice red crop milk, with mom's head a convenient spoon.

As for that one-legged posture, it could be that this enables flamingos to conserve body heat, or perhaps it's a behavior that uses less energy. What is known about flamingo legs is that the joint that appears to be the knee is actually the ankle. Where is the knee then? Up under their feathery bodies? Puzzling.

It is likely our visiting flamingos will head to warmer temperatures once winter arrives. Although they don't migrate, they will relocate as comfort dictates. Some avid birders hold out hope that they, and the other displaced American flamingos, will decide to make Florida their permanent home.

So, is anyone relating to a flamingo today? Feeling a bit blown off course? On your last leg? As for me, my relatable animal is a wolf. Always a wolf. Why? That's fodder for another column, another day.

A Placental Proposal
September 21, 2023

I just stumbled across some excellent news for the male specimens out there. It turns out that despite the ability to clone mammals from females, we still need the father's genetic contribution to ensure a healthy placenta. And healthy placentas are all the rage, biologically speaking.

I wish I'd known this yesterday, when Joe and I ran across a friend out walking the dogs at Kiesel Park. He was all aglow because his first child is due this November.

Mark is an adventurous type. "Oh, this will be your greatest adventure yet," I assured him. And then, yet again, words just fell out of my mouth, direct from some strange recess in my brain. "Wait till you learn about the placenta."

"I can't wait," he said eagerly. We parted ways, both of us now pondering placentas.

Before I begin extolling the wonders of this most amazing organ, let's establish the fatherly bit. I just learned that the placenta is not a maternal organ, but that it develops from instructions from both parents' DNA. That makes sense, I think, sipping coffee, now thinking of the placenta as something arising from the fertilized egg and not a magical growth from the mother.

But what's this? It's the paternal genes that dominate this wonderous organ.

Think of the placenta as having two sides, one attached to the uterine wall of the mother and the other attached to the developing offspring. It turns out that due to a phenomenon called imprinting, in which one of a pair of genes is silenced, the embryonic side of the placenta has only the father's genes.

Interesting, I think. Then I remember that with cloning, a process using a spark of electricity to stimulate embryonic development, there are no genes from the father. Where is the paternal placental input in the case of cloning?

It turns out that there are problems upon problems in the placentas of cloned animals, described in piles of scientific papers, written in words almost indecipherable to most of us, but there all the same.

Placental problems have been linked to health issues in adults, including asthma, coronary heart disease, and diabetes. There you have it. Cloning will never eliminate the need for men.

Speaking of men, I know that we're right in the midst of football season, and that you may not have awakened with placentas on your mind, but hear me out. The placenta deserves at least as much attention as the SEC.

Recall from high school biology that the placenta plays several important roles in fetal development. For one, it acts as a lung, transporting oxygen to and carbon dioxide away from the fetus, with the barrier between the mother's and child's blood so thin these molecules can cross over, while still ensuring the blood never mixes. For another, the placenta acts as a liver, metabolizing nutrients for the developing child and protecting it from harmful substances. In mechanisms still not completely understood, the placenta works in tandem with the immune system, protecting the fetus from attack by the mother's immune system while delivering immunities to the developing child from the mother.

That complicated bit about immunities sent me down a convoluted research rabbit hole from which I emerged only more baffled, and impressed. If any immunologists want to clarify the role of the placenta in protecting from and providing immunity in fetal development, please write a guest column.

What you didn't learn in high school biology was that the placenta, dubbed "the wild west of the human genome" by Sam

Behjati, a pediatric scientist specializing in cellular genetics and cancer, is a mosaic. A tapestry. A confusing quilt.

Placental and fetal development begin differentiating as early as the first few cell divisions. Behjati and his team discovered that placentas are rife with genetic abnormalities not found in their fetuses. Could it be that the placenta is the trash-can for mutations which could harm the fetus? Or is the placenta an untamed garden of genes, while the fetus is kept, pruned of potential troublemakers? The first study of the human placenta's genome, published in *Nature* (March 2021), leaves us with these questions.

Regardless, the ability of the placenta to survive as a foreign agent in a body, and more interestingly, to stop growing when the time comes, is fuel for research on cancer cures and prevention.

All this science aside, think about it. If not for placentas, likely triggered by a retrovirus some 60 million years ago, fetal development would only occur in eggs (or some unimaginable alternative). There'd be no mammals. There'd be no cuddly babies smiling to repeated readings of *Goodnight Moon*. No dogs, no horses, no meercats. No football, no SEC.

I propose we all stop and ponder the development of the placenta, and toast to the mysterious work of this live-fast-die-young temporary organ, this under-appreciated multi-talented mass of spongy tissue. And to all the male mammals, let's also raise a glass to thank you for your contributions to the healthy placenta.

A Whopping Long Line of Marys
September 29, 2023

Once I brought my college roommate, named Mary, to a family reunion in Panama City. Countless cousins smiled at her and asked, "Now how are you related exactly?"

It's easy to confuse. I come from a whopping long line of Marys.

My great grandmother was a Mary. She had a sister-in-law Mary, a daughter Mary, four granddaughters named Mary, and two great-granddaughters named Mary. My mother, not a Mary, told me that she decided to name me Mary because if the name was good enough for Jesus's mother, it was good enough for her baby girl, but family history suggests otherwise.

I've got all these Marys and other relatives on my mind today because last week, my beautiful Great Aunt Pat, age 95, the youngest of the ten siblings which included my grandmother Mary, died.

It almost defies belief that the last of these siblings is gone. They were around for a long, long time. While great uncles died at ages 48 and 64, the other two made it to 90 and 91. My great aunts and my grandmother lived to be 94, 95, 97, 98, 99, and 99. At one time there were eight siblings over the age of 80.

When Robert and I were little, we used to ride our bikes over to visit our Great Grandmama Mary, who we just called Grandmama, and Aunt Frank. They lived in a brick house with a giant Magnolia tree in front, a house which still stands on Dumas Drive. My mother told me it wasn't unusual for there to be a designated spinster in the family back in those days, someone to take care of her parents forever. I suppose that was my Aunt Frank's lot.

While Grandmama was spun silver and sugar, with long white hair which she braided and coiled into huge loops on her head, Aunt Frank was persimmon and vinegar. She used to shoo Robert and me out of the house to go play in the yard, and she told my father to tell us to stop bringing that dog, Bowzer, with us whenever we visited. We minded her when it came to playing in the yard, but we never, ever, ever told Bowzer he wasn't welcome.

Later, possibly after Grandmama died, Aunt Frank joined the ranks of silver and sugar, and like the rest of her siblings, my memories of her are salt of the earth, fun-loving, storytelling, fly-swatting, rocking-chair-rocking, sweet-tea-and-sometimes-gin-and-tonic-sipping, starshine summertime delight.

It was always fun when the relatives came to Auburn. Aunt Frank's house filled up with folks. We never knew how we were all related, and we drew our own lines. Adults who stood around and talked were aunts and uncles. Everyone who went outside and climbed trees was a cousin. These lines are hilariously criss-crossed, with nieces and nephews older than aunts and uncles, and cousins spanning multiple decades.

(Recently, I bought my girl cousins Amy and Katie "Cousin Explainer Tea Towels." According to the tea towels, Amy and Katie are my first cousins once removed, and their kids, who I call nieces and nephews, are my second cousins.)

The best of times were the summers we spent at the Sun-N-Swim, a 32 room, two story motel right on the water at Panama City Beach. We'd pile into a handful of rooms which would take on designations. There was the card playing and game room, the music room in which Uncle John could be found jamming on the accordion, the bar room where we youngsters were allowed to mix up Bloody Marys and take them into the ocean, and the quiet rooms for reading and rest.

We cousins ran amok, sneaking cigarettes and beer, getting sunburned, and going to the movies. And of course, we floated endlessly in the ocean on our blow-up rafts, dreaming about cute members of the opposite sex, and where we'd go to college someday.

The cousins who lived in Panama City surfed and excelled at spray paint art, completely exotic talents to us land-bound kids.

A few years ago, I graduated from a traditional bathing suit to long sleeves, shorts, and a big, wide hat. I thought about my grandmother, all those great aunts, and my mama and her sisters back in Panama City, laughing, smoking cigarettes, drinking high balls, walking on the beach looking for shells, loving all us kids, playing poker, and singing. In my memory they were dressed like I am now, comfortable clothes, never having to adjust their swimsuits or worry about tan lines.

Those women are my heroes, my fashion icons, my ancestral guiding stars. They are the spark passed on to me in my mitochondrial DNA.

Learning Without Brains
October 5, 2023

"Brainless Brilliance: Jellyfish Stun Scientists with Learning Skills." The headline from SciTechDaily.com is bold and loud on the top of my personally tailored news app. The little fairies that live in my computer know me well. I'm clicking. As with slime molds, I'm intrigued by brainless organisms who challenge our ideas about learning. And no, I'm not talking about politics.

The jellyfish de rigueur is the Caribbean box jellyfish, *Tripedalia cystophora*. The Caribbean box jellyfish is one of about 50 species of the invertebrate class Cubozoa, jellyfish with a cube-shaped medusa. Its relative, *Chironex fleckeri*, nicknamed sea wasp, is renowned for being the most dangerous marine animal, its venom inducing cardiac arrest within minutes.

I grew concerned as I began reading about the Caribbean box jellyfish hanging around in Florida. Of all the things to be worried about while swimming in the ocean, box jellyfish were not on my list. Turns out they needn't be, unless I'm swimming off the northern coast of Australia where the sea wasps live. These Caribbean boxies aren't harmful to us, as far as we know. And they are tiny, like a fingernail.

My coffee tastes a bit better knowing that the brainless yet smart jellies aren't out to kill me.

The fact that they are now permanent residents in Florida and our mid-Atlantic shores, a move triggered by the gradual warming of our oceans, is still concerning as I ponder the health of the seas. But for now, this sweet October morning, I'll put worry to the side and welcome wonder and awe.

Let's get interactive. Grab a piece of paper and draw your current mental image of a box jellyfish. I'll wait. I'll even give you doodle space.

Perhaps you drew a square-ish medusa, given their name. Did you draw the flat stomach in there, comprised of four gastric pouches, inside of which are the butterfly shaped gonads? Did you cluster up to 60 tentacles into four groups, spaced evenly around the square body? And be honest. Did you draw the 24 eyes along the rim of the medusa, with four of them always peering upwards?

24 eyes. I can tell you exactly where I was when I learned that box jellyfish have eyes.

It's been known for some time that these brainless jellyfish have some kind of primitive sight, though exactly how it works remains mysterious.

The Caribbean species has made the news as they appear to be able not only to employ their tiny eyes, but to learn. Researchers placed their subjects in an aquarium adorned with models of mangrove roots. At first the little jellies swam close to these fake roots, bumping into them frequently. But over time they began avoiding them, increasing their distances, and cutting their contact with them by 50%. Learning.

Because some folks spend their days researching these things and not walking in circles looking for their car keys like the rest of us, they have identified "visual sensory centers" called rhopalia, in the jellyfish. Each rhopalium houses six eyes. Through prodding and electrically stimulating and sensing and observing and all manner of scientific doings, they determined that there's some combination of visual and mechanical stimulation responsible for this associative learning.

More research is in the works. If you're thinking this sounds like an awfully big investment in a tiny brainless blob, consider that it's not just about the jellyfish.

To quote SciTechDaily.com (9/24/23): "It's surprising how fast these animals learn; it's about the same pace as advanced animals are doing," says Anders Garm (senior author of the published research). "Even the simplest nervous system seems to be able to do advanced learning, and this might turn out to be an extremely fundamental cellular mechanism invented at the dawn of the evolution of the nervous system."

Before we leave this morning's ponderings, here are a few words on the Australian box jellyfish.

Australian box jellies are one of the world's deadliest creatures. Their venom directly attacks the central nervous system. It's said that some victims die before they can reach the shore.

To avoid such stings, swimmers in areas where these deadly box jellies are found might wear special sting-resistant SCUBA suits, or swim in areas surrounded by specially designed nets. Some studies suggest box jellies do not like the color red, prompting swimmers to wear red suits and employ red nets.

Oh, you sea wasp, with half a million stinging darts on each tentacle, each dart with enough venom to theoretically kill up to 60 people, so sensitive that a mere brush can set off your stinging, swimming around in the ocean somehow seeing with your 24 eyes, you have my respect. And yes, I said swimming, not just drifting like most jellyfish do.

From slime molds to jellyfish, the realization that some brainless organisms can learn only reinforces what I know to be true about the living world: It's endlessly more complicated than we can possibly understand, and that all living things are worthy of our respect.

Pinecone Pets
October 19, 2023

If I'm not careful, I'm going to become one of those crazy pinecone ladies. You read that right. I have no inclination to collect cats. They're hard on the small critters and birds I love so. Instead, I've started collecting pinecones and rehoming them around a mulberry tree in the yard.

Not only that, I've caught myself referring to them as pets. Yikes.

You see, they're not just any pinecones I collect, but longleaf pinecones, the largest cones of any of our southern pine trees, some as long as 12 inches. I used to put them around the house as decorations. One day I realized I had reached the sane limit to pinecone décor and tossed my daily collection off the deck, where they landed under that mulberry tree.

A few days later, I noticed they had changed. My once wide open, triangular shaped pinecones had closed up. Now they looked like little grenades, all tightly bundled and dense. Weird, I thought, and promptly forgot about them until a week or so later when lo and behold, they were back to their wide open, spectacular selves.

And lo, I was hooked. All this time, I thought I knew a pinecone.

I grew up surrounded by pine trees. As children, we knew every loblolly pine on our one-acre lot, their bark patterns as distinctive as fingerprints. We knew sharp scent of the sappy scars, the scraggly branches, and the root patterns of every tree. We spent Saturdays raking the yard, hauling all the pine straw to a pile at the end of the driveway. I whiled away hours sitting in a pine straw nest, pretending to be a giant crow.

And oh, we waged glorious pinecone wars. We'd separate into armies, amass gargantuan piles of pinecones, and declare battles. Metal trash-can lids were excellent shields, and the army helmet liners my brother and his friend Chris found in a rubbish pile (so they said) protected our young heads.

Pine trees and their accessories were as much a part of my lived experience as air.

How is it that I never noticed pinecones were shapeshifters, specifically hygromorphs (objects which change shape according to humidity) until my 50's?

There's so much to know about a pinecone.

Recall from grammar school that pinecones have a sex. The female cones are what we think of as pinecones. The males, which provide the pollen, are smaller, pliable, and what my granddaughters call "worms." We used to mix these male cones in water, delighting in the brilliant yellow liquid resulting. I can only imagine the pressure on our little immune systems. To this day I'm not allergic to pollen.

Those wee pollen grains find their way into tiny crevices between the scales of the female pinecones while they are still on the tree, where they fertilize the ovules within. Those fertilized pinecones then close up tightly and remain on the tree, nourishing their developing seeds, for one to ten years.

Biological form and function are on full display here. The scales of a female pinecone, gorgeously twirled in perfect Fibonacci spirals, are perfectly adapted to keep the seeds safe, their primary duty, from prying critters and other elements.

When the seeds are ready, the female cone responds to the optimal conditions for opening her scales and releasing her treasures: warm temperatures for germination and dry air for best dispersal by wind. Wet seeds would fall straight to the ground, thus overcrowding the growing field. Therefore, when the humidity is high, her scales will close back up.

When the female cone has released her seeds, she drops to the ground. Of course, some cones fall prior to this due to wind, scrambling squirrels, or other factors. Those same squirrels have a field day shredding the scales off the pinecones, searching for the seeds within. Chomp.

Pinecones can go through several cycles of this opening and closing action, even after seed dispersal. The first bunch I tossed under the mulberry tree, well over a year ago, are still opening and closing with the weather, this ongoing flex predicable and oddly satisfying.

The mechanism for this action is due to a specific arrangement of the fibers cellulose, which swells in humidity, and lignin, which doesn't.

Folks' capacity for awe is varied and individual, and the fact that my pet longleaf pinecones keep opening and closing, day after day, with changes in the weather, may not float everyone's boat. So, let's consider the usefulness of this phenomenon.

Research is underway to manufacture smart buildings, inspired by the opening and closing mechanisms of pinecones to conserve energy, as well as creating a fabric which senses and responds to humidity. Google "hygromorph biocomposites" to learn more if inclined, and be amazed at biomimicry.

For now, I enjoy the low-tech, humidity sensing pet pinecones in my yard as a reminder that nature is crafty and clever, and to marvel that in a hundred lifetimes, I could never have imagined so fantastical a structure as the ubiquitous female pinecone.

The Barefoot Hero's Journey
October 2, 2023

Ah, the world remains as mysterious as ever, human behavior included. This is why my writers' group, the Mystic Order of East Alabama Fiction Writers, decided to devote this month's podcast to offering advice in the spirit of Dear Abby. I cannot reveal the content of the podcast before it airs, but I'm compelled to explore one wee bit of it now. I can't stop myself.

Our self-anointed Mystic Queen, Gail, told us about her latest newfangled contraption: a grounding mat. According to Gail, you plug it into the ground hole of a wall socket and drape it over your lap to harness Earth's natural energy and improve your well-being. She suspects it is working, but is unsure if it's a placebo effect. Gail, energized and sizzling, further explained that it mimics the benefits of walking barefooted outside.

I don't need one. I go barefooted. My feet, heavily calloused, look so bad I'm embarrassed to get a pedicure, but that's a small price to pay for being about to walk through my wooded yard, around my gravelly driveway, and down my rough paved road without the hassle of putting on shoes.

My tough feet have always been a point of personal pride. When we were youngsters, my brother Robert and I used to sew our initials into our heels with my mother's sewing needles and crimson threads.

In high school, I was a good student and didn't cause much trouble. I kept my grades up, and because I worked in the office I could finagle skipping school without getting caught. I only had two conversations with the assistant principal, Mr. Robinson, about my behavior. One was a reprimand for sitting in the parking lot with my boyfriend during lunch. The other was a warning about my bare feet.

"If you come to school one more time with nothing on those feet, I'm sending you home," he said sternly.

The next day I showed up with strings wrapped around my big toes and tied behind my ankles. He was so amused he did not send me home. What a little twit I was!

But enough about my feet. What's this about plugging a blanket into a socket and grounding yourself?

I suggest you skip that big cup of coffee now because I'm about to toss around some jargon which might cause you to feel queasy, perhaps evoking a gag. You may want an ice-cold Coca-Cola instead.

To understand this grounding phenomenon, we must first talk about earthing (you were warned), which is the practice of walking on Earth without shoes. Also called barefoot walking, it is different from just walking barefooted according to the earthing gurus, though the difference is never explained.

Aside from the orthopedic perks of strengthening arches, using muscles in your ankles and calves that aren't used when shod, and the sheer delights of cool grass underfoot and squishy mud between one's toes, there are some electrical perks as well. Indeed, there is growing evidence that our overall health could benefit if we'd take off our shoes outside.

The gist is that Earth's surface has an overall negative electrical charge. The National Institute of Health hypothesizes that when we are in contact with Earth, Earth's free electrons spread into our bodies where they create antioxidant effects around damaged tissue and prevent or resolve inflammation.

In short, wallowing about on the surface of our planet might improve sleep, better our moods, help us heal from injuries, and generally assist "the very fabric of our body."

Only now, wallowing about is called earthing.

It is the duty of each generation to stand in the yard, barefooted or not, and shake a metaphorical fist at the youngsters of the present, grumble about "kids these days," and begin sentences with, "Back in my day." I have politely asked my family to muzzle me if I start doing that, so it's with careful and conscientious deliberation that I continue.

But y'all. There are articles all over the internet on "How to Earth." There are websites that will sell you classes on how to walk around barefooted or, excuse me, how to barefoot walk. There are products galore, including "Grounding Shoes." Are these like the strings I wore to school? Not quite. Available in models from sandals to sneakers and even boots, they have genuine copper conductors built into the soles. The pair of sandals I clicked on cost $215.00.

I'm happy for all those who can no longer walk barefooted outside that there are grounding mats and even shoes to help their bodies reconnect with the electrical matrix we are meant to dwell in. As for me, I'm headed outside now to lie down on the leaves with my dogs.

The hero's journey can be tangled, twisted, arduous, and exhausting. Isn't it grand that ours has landed us right back where we were as children, wallowing about on the ground and running around barefooted?

Chills, Goosebumps, and Talking Heads
November 2, 2023

I learned a new word today: frisson. In short, it's the tingling feeling you get when moved by art. Frisson (FREE-son) might cause your skin to break into goose bumps; chills to run along your spine; and your soul to soar to the heights of ecstasy, plummet to the depths of despair, or both. Dopamine is involved. Also known as aesthetic chills and psychogenic shivers, frisson is triggered by a variety of expressions including music, art, nature, and films. It can also be stimulated by eloquent speeches, and the practice of science and math.

Well knock me over with a 715-song Spotify playlist, curated by a team of neuroscientists, guaranteed to give you the chills and the shivers!

Theories abound, especially musical, on what causes frisson. Some say it happens when we expect one thing and are confronted with an unexpected sound, pitch, color, or key. Ethnomusicologists wax on about the circle of fifths and tonic chords, but here I am lost. Perhaps Emma, our daughter who is a professional musician, can fill us in.

New word in tow, my husband Joe and I set out on a date Saturday to see the newly restored version of Talking Heads' *Stop Making Sense*, described by many critics as the best concert movie ever made. In case you've forgotten, think David Byrne in his "Big Suit."

Settled in the Capri Theater, pleased at the decorous red curtains parting to reveal the screen, I did the math. The original movie was released in October 1984. I was living in Boston then and saw the movie at the Harvard Square Theater with my mother and stepfather. I was 20 years old, my mother 44. She had a huge crush on David Byrne and is reported to have tried to call him on the phone a time or two.

Watching David Byrne's eccentric, jerky head motions as he sang and danced, I pondered the changes in my own life and my mother's since that day we watched this movie together. She grew quiet and somewhat reclusive, and began each morning painting in her studio, blasting Mozart in her headphones so loud she damaged her hearing. Ever the music fan.

I remembered the summer we vacationed together in New Hampshire in a cozy wooden house on the banks of Gilmore Pond. Talking Heads had just released a new album, *Little Creatures*, which we played on an endless loop out on the porch overlooking the pond. That was the scene when my mother, celebrated watercolorist, started painting.

Joe and I got married at that same wooden house in New Hampshire two years later. I glanced at him, my husband of 36 years. Our marriage spans almost as long a stretch as between the original *Stop Making Sense* and this restored version we were watching now.

Amazing, this bookended space of almost 40 years, in which we have created a whole life we couldn't have dreamed of then, replete with grown children and two grandchildren!

And suddenly, amidst all these memories, as David Byrne sang "Burning Down the House," I got the goosebumps. Interesting. This is not my favorite of their songs. I don't have a strong emotional attachment to it, though I sang it at top volume in my younger days. But here it was, frisson in full display.

The goosebumps came and went throughout the song. When it ended, the audience, six songs in, broke into the first round of applause.

Well, I'll be. They frissoned too.

I bet this is why everyone stands for the Hallelujah Chorus of Handel's *Messiah*, I thought, King George included. Just thinking of those Hallelujahs gives me the shivers.

Throughout the rest of *Stop Making Sense*, I experienced aesthetic chills several times. I was in no control of when they started and stopped.

In a study published in the journal *Social and Cognitive and Affective Neuroscience*, researchers found that frisson-prone folk, that's between 55-85% of us, have more fibers connecting the auditory cortex to the emotional centers of the brain. I'm curious about those who do not experience these tingles, unable to imagine a frisson-free life.

Perhaps more fascinating than my skin's reaction to *Stop Making Sense* is the brain-blasting journey I took during that hour and a half, almost 40 years crammed into 90 minutes, every second slammed with disbelief that my mother is no longer living mashed up with overwhelming happiness that for 82 years, she was.

What was my brain doing? Researchers suggest an evolutionary advantage to frisson, helping us find connections and meaning between ourselves and our environment, thus strengthening our species. Neurobiologists have long established the close connections between music and memory.

Whatever it was that caused my emotional response, this evening with Joe, awash in memories of the last 40 years, accompanied by frequent chills down my spine as we watched perhaps the greatest concert movie of all time, made for a perfect date.

To Come Back as a Buzzard
November 9, 2013

"If there is such a thing as reincarnation," my father proclaimed during one of our deep, philosophical conversations, "I want to come back as a buzzard."

Upon recounting this to my mother, his ex-wife, she replied, "Once was not enough?"

My father was tickled and honored with her wry humor.

This came back to me when, shortly after his death, we were on our way to his memorial service when there, on the grassy area before the driveway to Bethany House, stood a flock of buzzards. I never saw them congregate there before or since. Mystical? Coincidence? Bob Mount reincarnated? Who knows why they flocked to the place where he died on the very day he was memorialized.

As for his wish, you might ask why on earth would someone want to come back as a buzzard, oft reviled, misunderstood, and largely unappreciated. Turns out some folks appreciate them. As obligate scavengers, vultures' uncanny ability to locate corpses and to rid the land of carrion, including diseases possibly harbored in rotting flesh, led them to be associated with divination and purification in some cultures.

Vultures were sacred to ancient Egyptians, often represented as gods. In some Greek and Roman myths, vultures were seen as talismans of happiness. They were considered auspicious in ancient Rome and have a part in the story of Romulus and Remus, legendary founders of the city.

Today's scientists assess a value of $16,000 per bird for their contributions to maintaining the cleanliness of our land, water, and air.

Maybe we should all strive to come back as buzzards.

I can hear some of you now, shaking your head and tsk-tsking. These birds I speak of, they aren't really buzzards. Called "buzzards" by early settlers, who mistook them for buzzards of the Old World, the nickname stuck. What we buzzards are technically vultures.

Vultures come in two categories: Old World and New World. As well as habitat, differences include urohydrosis, that is, the act of cooling oneself through defecating down the legs, something only New World vultures do, and talon strength. New World vultures' talons are relatively weak, so weak that they cannot grip and carry food to their young. Instead, they feed their nestlings via regurgitation.

For a while, taxonomists classified Old and New World vultures together. New evidence reveals they aren't so closely connected after all. New World vultures are more related to storks, while Old World vultures are more closely related to eagles.

This is an example of convergent evolution, that is, different organisms adapting in the same way to similar surroundings and circumstances, but not necessarily sharing the same family branches.

Here in North America, we have three vultures: the California condor, the turkey vulture, and the black vulture. In Alabama, those enormous, carrion-cleaning, pine-tree-roosting birds we see everywhere are mostly black vultures. Though the population of turkey vultures is larger, black vultures congregate closer to humans.

You have to hand it to black vultures. Compared to their turkey vulture friends, their sense of smell is a bit dim. They keep a watchful eye on turkey vultures, who can smell a dead animal a mile away, and join in the journey to the buffet. Although stronger one on one, turkey vultures are usually outnumbered by their socially advanced black vulture friends, and though they do the work of locating the meal, they often must wait their turn to eat.

You can tell our buzzards apart (with all due respect, I cannot stop calling them buzzards) by the colors of their heads. Turkey buzzards have bright red heads, and black buzzards have black heads. Both birds' heads are featherless, all the better for plunging deep into rotting flesh and coming out clean as a toothpick in your grandmama's pound cake.

I've had the pleasure of knowing a few buzzards up close and personally. One of my dad's friends, Dan Speak, had a buzzard named Bobo. Bobo was a human imprint, likely by design. He followed his human around like a dog, accompanying us on woods walks, fetching sticks and playing tug of war.

My father was green with envy for want of a pet buzzard. Not content to have had a crow, he longed for a vulture to call his own. He somehow ended up with one. Though not a human imprint, it did enjoy our company and loved a good scratch on the bald noggin.

I'm reminded of my husband Joe's first trip to Alabama from his home in Massachusetts. As if the baseball hat commanding, "Keep Alabama Beautiful: Yankee Go Home" we saw at the gas station wasn't enough, as soon as we reached the house, Daddy's giant buzzard hopped over to greet him, demanding attention. Joe passed this make-or-break moment with flying colors, delighted by this friendly bird named Buzz.

This weekend while horseback riding with three friends, we came upon a large flock of black vultures. Even the horses were impressed with their wingspans. We stopped, we ooooh'ed, and we aaaah'ed, all of us fans of these hard-working, intelligent birds.

And my heart was warmed, knowing that somehow, the spirit of my dad was among us.

Wild Turkeys: A Success Story
August 16, 2023

"Mom," I heard Sarah, our daughter who was visiting last weekend, scream. "Get out here! There's a dinosaur in the yard!"

Sure enough, our neighborhood turkey was prancing unawares up the driveway toward the house. She wasn't bothered a'tall with the yapping dogs or the cacophonous clatter when I dragged the big trash can down the gravel drive.

"Hello, Henrietta," I said. She ignored me.

Sarah gaped, dumbstruck, as this long-legged, prehistoric beast strutted past her into the trees.

It seems this turkey, known as Henrietta to some and Maude to others, has been here since we moved to this neighborhood 24 years ago. In fact, wild turkeys only live for three to five years, so I assume we have a regular run of wild turkeys living amongst us.

Turkeys, like other displaced wild animals, will indeed take up in neighborhoods. There are reports of wild turkeys losing their fear of humans, even becoming aggressive and damaging property. Still, in my experience, our Henriettas are peaceful and polite. We welcome her/them. Not only is there a mystical feeling of connection when we come across wild animals in our midst, but one wild turkey can eat up to 200 ticks in a day.

Henrietta's survival is a testament to the wise ways of wild turkeys, who can fly in bursts of up to 55 mph and roost in trees. Wild turkeys form strong social bonds, show affection to one another, and have been seen putting their own self-interest behind that of other turkeys. Adaptable, courageous, and exhibiting a wide range of personalities, these turkeys are far from the mythical dumb beasts who drown in the rain.

Perhaps domesticated turkeys gave wild turkeys their ill-earned reputation. Domesticated turkeys have every wild instinct bred out of them and can barely walk for the Frankenstein-like selection of meaty genes that factory farms have embraced. It's illegal to release domesticated turkeys for fear they will contaminate the wild turkey population, although interbreeding is unlikely. Domesticated turkeys are so malformed they have to be artificially inseminated.

There are six species of wild turkeys, varying in gobble strength, size, and plumage. I highly recommend pouring a cup of coffee and googling images of the Ocellated Turkey. Be amazed at its peacock-esque shades of turquoise, orange, and green. These exquisite turkeys live only on the Yucatan peninsula. Sadly, and not surprisingly, the Ocellated Turkey is ranked as Near Threatened due to overhunting and habitat loss.

What may surprise you is that our own wild turkeys had their bout with near-extinction. When the settlers arrived, millions of turkeys roamed across the land. Hunting and habitat destruction led to such a decline that by the early 1900s, folks were worried they'd go the way of the dodo bird.

By the 1930s, there were as few wild turkeys here as polar bears now. The passenger pigeon had gone extinct, and our bison population plummeted. Rather than shrugging shoulders and declaring the loss of wildlife collateral damage of "progress," the American conservation movement kicked into high gear. Through the enactment of regulations, protections, game laws, and other policy efforts, wild turkeys began a recovery.

Mother Nature helped too. As desperate families abandoned their failing farms, native grasses, trees, and shrubs returned, a welcome respite for struggling turkeys.

Another boon to our wild turkey population was the passing of the Pittman-Robertson Act, or the Federal Aid in Wildlife Restoration Act of 1937. Originally intended to promote preservation and restoration of life habitats of all species of wild, free-ranging fauna, this act was restricted in 1956 to include only game. Whatever its intended and enacted purposes, wild turkeys continue to benefit from this act, which distributes federal funds to states in proportion to the size of the state and its number of licensed hunters.

The restoration of our wild turkeys is considered by some to be our nation's greatest achievement in wildlife conservation. Today, over seven million wild turkeys inhabit our contiguous states, some wandering from forests into our backyards.

This year, while gathered at the Thanksgiving table with family and friends, I'll be giving thanks for the success of our wild turkeys, and pinning hopes on the wisdom of future generations, the passing of the Restoring American Wildlife Act, and an enlightened state of humanity capable of living in harmony with the wild, wild world.

Checking in With Gratitude
November 23, 2023

I've been staring over the top of this computer for some time now, composing and decomposing ideas about Thanksgiving, watching the woodsy world outside my window appear every so slowly at dawn's first light. First came the browns, then the greens, and oh my stars, here come the yellows.

I've pondered the Thanksgiving I'll have this year with family and friends on Edisto Island, I've explored the History Channel's delightful presentation on the first Thanksgiving, and I've googled images of cranberry flowers to see if they really look like cranes, hence the name.

I've filled my coffee cup twice, reached for an oversized denim shirt not because I'm cold but for the coziness of it, and found relaxation in the steady ticking of the clock in the kitchen.

Imagine the many unseen hands that did the work that allows me to enjoy this sweet, early morning reverie.

Who invented glass? What would the world be like without windows? And glory be to the original owner of this parcel of land for not selling the back lot, allowing us to have trees, those wonders of wonders, to admire through these windows.

Oh, the struggles to build a bridge to a barrier island back in the day that I can now drive across to reach Edisto. Oh, the humans upon whose backs the roads, churches, and homes were built that I might vacation there. Imagine the persistent and costly determination that has demanded environmental regulations to keep Edisto's wetlands, flora, and fauna protected from greed.

How on Earth can computer engineers figure out how to build computers that folks can then clickety-clack upon and see actual images of the first Macy's Day Parade, then with a swipe of a finger, gaze upon photographs of several species of cranberry flowers, all the while sipping coffee ground from a bean grown on a farm in Indonesia, roasted in New Orleans, shipped to Publix in Auburn, AL, and brewed in my kitchen?

Did I mention the marvel of the running water that came right out of my faucet, on command, that I used to fill the coffee pot this morning? What a miracle that with the mere twist of a wrist I can have clean water cascade down a stainless-steel portal onto my hands, splish-splash, willy-nilly.

I have so much to be thankful for, including permission to leave the preposition at the end of that sentence, which is technically okay if putting it elsewhere in the sentence makes things awkward.

As most of us know, practicing gratitude is good for us, just like our grandmamas said. Even the wisdom of our ancestors, though, gets fact-checked.

A groundbreaking study in 2003 by Dr. Rober Emmons, now considered the world's leading expert on gratitude, and Dr. Michael McCollough found that gratitude improved psychological well-being, thus sparking interest in and inspiring further research on the topic.

Current research reveals that practicing gratitude can improve sleep, lessen anxiety, lower diastolic blood pressure, calm the nervous system, and possibly reduce depression. Scientists are still trying to tease the correlation/causation relationship between gratitude and depression. Are people less depressed because they practice gratitude, or do people practice gratitude because they are less depressed?

Every November, we are besieged with reminders to be grateful for our blessings. Research suggests we should extend our monthly practice to daily.

I'm wondering if my morning musings count as practicing gratitude. Let's do a gratitude check.

First, what is gratitude? Gratitude is loosely defined as acknowledging the goodness in your life and the people or powers that made this goodness possible. To reap the benefits of gratitude, one must focus beyond oneself. I get a check here.

Next, it turns out that sitting around feeling thankful doesn't get the job done. We have to actually practice gratitude. Suggestions include writing it down (check), hitting pause to really think about what you are thankful for when you say "thanks" (check), redirecting your thoughts (check), and sharing your gratitude.

Sharing your gratitude does not mean writing about it. It means sharing it with the people or powers you are acknowledging. As of now, I have not earned a check mark here.

But wait.

There's still time.

Thank you, readers, for taking the time to read my columns. Thank you for the kind messages you've sent through friends, emails, and messages. I appreciate your encouragement and I love the many stories you've shared with me. Thank you, Auburn Villager and Carolyn Eddins, for giving my columns a home. Thank you, Brian Woodham, for being patient and precise. Thank you, big old world, for providing me an abundance of topics.

Happy Thanksgiving! I hope you all enjoy feasting, family, and effortless access to clean and abundant running water.

The Kleptoparasites in Your Oysters
November 30, 2023

I was eating oysters with my stepmom Janie, no, let me rephrase that, I was watching Janie eat oysters last week when lo and behold, she held a shell out to me to show me a wee, fat, mucousy crab nestled with the poor oyster she was about to consume.

As if eating a raw oyster isn't enough, she had to go and find a little crab living in it. Janie plopped the crab onto a napkin and I watched, horrified and fascinated, as its teeny legs unfolded as if it were still alive. But wait, its teeny legs weren't unfolding, they were waving, its pinchers were pinching, and, in fact, it was still alive!

She found about four little crabs in her dozen oysters that day. I ate my side of macaroni and cheese and washed it down with two beers, wishing it were ten. I tried to pretend that I wasn't thoroughly, undeniably, girly-girl squeamishly, knock-me-over-with-a-monogrammed-feather grossed out, but I was.

"Look," Janie said, licking her fingers on one hand and holding her phone to me in the other. "They're pea crabs."

I leaned in. Sure enough. Delicacies. Happy little critters living in oysters, stealing their food, indicating a healthy ecosystem, and when chomped upon, bursting with a briny, sweet taste. It turned out she was lucky to find those crabs, prompting her to return the next day with a different set of family and friends who eagerly ate the live pea crabs along with the oysters

"You ate them ALIVE?" I asked.

"Well, the oysters are alive too until we whack them out of the shell. You knew that, right?" No, I did not know that. I'd have gone happily to my grave not knowing that.

Squeamishly intrigued, I set out to learn more about this little jewel of a stowaway.

Technically, there are two species of little kleptoparasites lurking in oysters: pea crabs (*Pennotheres ostreum*), and oyster crabs (*Zaops ostreus*). The species are similar, and both are commonly known as pea crabs.

Pea crabs prefer waters with a high salinity content, and are found in the Atlantic, with greatest concentrations along the Georgia/South Carolina coast. Spawning about a month later than oysters, free-swimming juvenile pea crabs cohabitate near oysters. The females take up residence in young oysters, nestling themselves in the oysters' gills, freeloading their meals. This does not harm the oysters, but if food supplies are low, it can cause the oysters' meat to be thin. The male pea crabs swim along externally fertilizing the females as the females roost in the oyster. Later, around 8000 baby crabs head for the waters to begin the life cycle anew. Unlike the soft-bodied females, males develop a hard shell.

If you're a raw (live) oyster fan and have never found a pea crab in your oysters, it could be because you are getting your oysters from the Gulf, or other areas where pea crabs aren't abundant. Additionally, most oysters are sold in restaurants, where chefs and shuckers toss the little tasties to the side so as not to gross out their ill-informed patrons.

I'll close this column with an introduction to an article published in *The New York Times*, 1907. "One of the sweetest and quaintest viands known to man is now so generally neglected that more than 50 per cent of the people who think they know something about good eating have never tasted the dish. This food is the oyster crab, the little soft-shelled creature of the sea, that seems to have all the sweetness and delicate soft savors of the entire crab family concentrated in its tiny body." This piece includes recipes, including the "daintiest" method of serving pea crabs which makes use of a dollop of mayonnaise tainted pink by beet juice.

Bottoms up, oyster eaters. Fear not the "redneck toothpick," so described because it squirms and tickles the mouth a bit before sliding down the gullet. I'll be over here looking the other way.

Hold Fast to Your Heart Horse
Mary Dansak, for Dec. 7, 2023

As of April, an amazing thing happened in my life. I reconnected with horses.

I was a classic "horse girl," my room postered from floor to ceiling with horses, my shelves full of Breyer horses, my parents exhausted and begrudgingly relenting to horseback riding lessons, and finally, getting a horse, Colti, for my 13th birthday. Tragically, Colti died nine months later, and although I immediately got another horse who I loved beyond measure and who lived to a ripe old age, I never got over the blow of the losing that first horse.

For reasons still unclear to me, I began ruminating on the death of Colti last spring and found myself unable to quit crying. Rather than stifling this sadness, which I'd done for over 40 years, I went embarrassingly public with it which resulted in the stars aligning to bring me back to horses. Oh universe, you strange and wonderful place!

Thanks to generous friends, I now have full access to horses. One horse in particular, a buttercream palomino Quarter Horse named Jasper, has my heart. In fact, the term among horse folk to describe a horse they love is "heart horse." Sometimes people recognize their heart horses immediately, other times it's a slow dawning.

Heart horse. How silly. But it's not.

Horses have very large hearts, five times larger than ours. When horses are among their equine friends, their heart rate variability (HRV), that is, the beat-to-beat changes in heart rate, synchronize.

Perhaps this is true of all herd animals. Interestingly, when they are near horses they don't know, this does not happen.

Research by Dr. Ellen Gehrke, consultant and professor of international business and management and general horse enthusiast found that horses' HRVs also mirror those of the humans they are in close contact with. Maybe this explains the tried-and-true adage that horses can sense our emotions, coupled with the advice to breathe slowly and deeply if we begin to feel nervous around them.

According to *Horse Sport Magazine*, Gehrke's research does not stand up to peer review, the gold standard for scientific research. However, other research found that when riders were forewarned about a potentially startling event on a trail ride, one that never actually existed, both the riders' and horses' heart rates spiked. This research did pass peer review.

On and on this human-heart-horse connection research goes, winding in and out of claims that humans and their horses share electromagnetic fields generated by the horses' enormous hearts, and other sciency-sounding attempts to explain just what this intense bond is.

Perhaps we will never fully understand why horses appeal so strongly to some of us. What we do have is clear evidence that for whatever reason, horses are therapeutic in many instances. To quote the American Hippotherapy Association, "Hippotherapy refers to how occupational therapy, physical therapy, and speech-language pathology professionals use evidence-based practice and clinical reasoning in the purposeful manipulation of equine movement as a therapy tool to engage sensory, neuromotor, and cognitive systems to promote functional outcomes. Best practice dictates that … professionals integrate hippotherapy into the patient's plan of care, along with other therapy tools and/or strategies."

All that aside, here's the thing. Ever since I left my horse in Alabama and went off to Boston for college, I had a nagging tug at my soul. This longing never once left me, that is until now. Being with Jasper and his friends soothes me in a way no other activity in my life can. Knowing that I can see him anytime relieves an inexplicable anxiety. Honestly, it's overwhelmingly grand.

I've watched people and their heart passions with wonder, being careful not to dismiss things I don't understand. Golf, for instance, appeals to me about as much as cleaning the tops of the kitchen counters. Hunting is out of the question. However, seeing the drastic measures others go to to golf or hunt assures me that there is something about those activities that's deeply compelling to their fans.

I recently asked some random people what they would love to do more if they had the time and ability. Most said travel. Some said hunt. Some would cook, golf, or read on the couch with their dogs. Some people had no answer and had never considered the question. They walked away with tilted heads and mysterious smiles, freshly pondering their fantasies.

I'm not one to be bossy, telling everyone to Live, Love, Laugh and such, but I have a strong recommendation. That is to remember, even when battered by the daily grind, what it is that you love. Keep your passions right in the front of your brain, let those dreams be noisy provocateurs.

Jasper brought me back to a place in my soul I almost forgot. To anyone feeling discombobulated by life's slings and arrows, I say, hold fast to your heart horse. Poster your walls.

Ode to Funchess Hall
December 14, 2023

When we were little, my brother Robert and I owned this town. We rode our bikes through all the neighborhoods, behind the houses, in every parking lot, through downtown, and all over campus. Bowzer, our faithful black and tan hound, accompanied us on all escapades, patiently waiting while we read the comic books in Toomer's Corner Drugstore, played pool in the basement of the Auburn Christian Student Center, or loaded up on paper bags full of candy at Zippy Mart.

We had many hangouts, but by far, our favorite was Funchess Hall, located on the south end of the university, across the street from the Heart of Auburn Motel. It didn't hurt that our dad had an office there, and taught classes in the building now nicknamed "Fungus Hall" by those who don't appreciate its funky aesthetic.

I loved Funchess Hall, the whole thing, from Charlotte, the friendly diamondback rattlesnake, to the stench of rotting sawdust and formaldehyde.

I loved hurling those huge glass doors open and running up the stairs. Halfway up the flight, a pair of mounted animals, a bear and a moose, hung from the wall. The moose was remarkably huge. My heart longed to see a real one, especially since my mama's nickname for me was Moose. I'd stand there on the landing and relate to the giant herbivore before me, leaning precariously across a small railing to be nearer.

The bear, on the other hand, scared me silly. Mouth open, huge teeth exposed, his lethal bear claws reached out for me. I stared, awestruck, and to this day, I maintain a healthy fear of being mauled to death by a bear.

Safely past the bear, I'd run up to the second floor where my dad's office was. Wait, I think it was the third floor, and that we entered on the second. Not only is my brain old and tired, Funchess Hall doesn't follow the typical rules of physics.

Whatever floor it was, I knew when I'd arrived because there was a great glass display case of various stuffed mammals and reptiles to greet me. I could watch, in real time, the raccoon's fur slowly disintegrated, and grainy sawdust leaked out of the gopher tortoise's shell.

After a brief hello to my animal friends, I'd look for my dad. He had an uncanny ability to whistle out of both sides of his mouth. Following the tell-tale human harmonica, I'd find him walking the halls or leaning in a doorway of someone's office, cigarette in hand. His blue eyes twinkled when he saw me, even when I caught him in the middle of teaching a class.

"Hey there, Little Sis," he'd say.

Then it was off to visit "the museum," a mystical room of tanks, cages, and buckets full to the brim of live specimens. I remember fondly sirens, hellbenders, millions of feeder mice, a gila monster, snapping turtles, box turtles, and lots and lots of snakes, the queen of which was Charlotte, an impossibly enormous diamondback rattlesnake.

At least I think she was a diamondback rattlesnake. Childhood is an unreliable narrator.

Charlotte has her own stories, those that my dad's graduate students, now grown men with gray hair, can tell. For me, she was just a big, beautiful rattlesnake. Charlotte was docile, having been in captivity for a long time. Robert and I would put our faces right up to hers, staring intently into her eyes through a glass pane. Sometimes her body would be pressed against the glass and we could examine the intricacy of her scales for long, mesmerizing minutes.

Despite her pit viper head, a shape that seems to draw fear in humans, we found her cute. She stared at us as we stared at her. Yes, we bonded with Charlotte. Every now and then she'd rattle, a noise now ingrained in my mind, one I've never once heard in the wild. How can she rattle so fast, I'd wonder, staring at the vibrating buttons on the end of her tail. She flicked her tongue in and out, the mystery only deepening.

Although we had total freedom to explore Funchess Hall, and all the buildings on campus for that matter, there was an exception: we were not allowed to ride the Funchess freight elevator by ourselves.

"That thing'll cut you in half," I can still hear my dad warning us. Yet every now and then, he'd take us on a ride.

We'd step into its great steel maw and Daddy would reach up and down at the same time, pulling on two opposite canvas straps that closed the doors horizontally. Then he'd reach over and yank a metal gate that expanded across the gap. We were in it for real now. He'd push a few buttons and the elevator would come to life, creaking and groaning. Slowly, slowly, it would descend to its destination. Daddy would reverse the door-opening process, and Robert and I would emerge on a new floor, wide-eyed in excitement, and beg to ride it again.

I haven't been in Funchess Hall in many, many years. It's time for me to pay a visit to the bear, the moose, and the ghost of Charlotte, who no doubt slithers the floors of Funchess Hall on quiet, dark nights.

Postscript: I did visit Funchess Hall after writing this column, my granddaughters Ruby and Annabelle in tow. While the old building directory, with its felt rows and white plastic push-in letters and numbers, still displayed the names of my dad and his long-gone co-workers, his office now belongs to someone else. The display of taxidermied animals is gone, likely to dust, and the museum of specimens was closed back in the 80s. That smell, though, the formaldehyde and must, still permeated the air, and lots of ghosts did indeed slither through the halls.

Ruby and Annabelle were enchanted by the moose and the bear as well.

Mistletoe
December 13, 2023

Why hello, winter solstice! The day has come at last, marking the exact moment when our home hemisphere is tilted as far back as it gets from the sun in its yearly lap around our star. We hunker down for this long night and enjoy all our holiday lights and evergreens, reminders that we haven't died in this darkness at all, and that light will return, year after year.

One of my favorite totems of rebirth in the darkness is that romantic, round leafed, white-berried, kissing-causing plant, mistletoe. No matter that it is a parasite, eventually sapping nutrients from its host tree so hungrily that it can kill the poor tree, or that its name translates to something like "poop twig." No matter that it's pearly white berries contain a goo so sticky it is used to make glue traps for birds. No matter that those berries, technically drupes, are mildly toxic, or that its mere presence can cause unwanted advances to be made on your personal body. No matter that its associated traditions are so steeped in paganism that it was shunned from Christian churches at one time.

All that aside, I am quite fond of mistletoe. This affection probably stems from, you guessed it, childhood.

Early entrepreneurs, Robert and I took advantage of the public's love of mistletoe at Christmastime when we were youngsters. We'd ride our bikes over to the pastures behind the Episcopal Church on Church Hill, ignore the cows who eyed us skeptically as we scooted between strands of barbed wire, and shoot clumps of mistletoe down from the trees with our BB guns.

Satisfied, we'd stuff our mistletoe cache into pillowcases, our answer to every toting need from collecting roadkill to candy at Halloween, and pedal home to sort our loot on the big wooden picnic table on the patio.

Mama would bring out the curling ribbon and the scissors, and we'd set to bundling our mistletoe into different sized clusters. Red ribbons always marked the largest sprigs, which we sold for twenty-five cents each, green ribbon was for the ten cent bundles, and we'd decorate the leftovers with white ribbon, more work than our price of five cents was worth if you forgot to monetize the fun.

Off we'd go, hauling our wares on a red wagon, with Bowzer, as usual, accompanying us.

I wonder what those neighbors thought when they saw us coming. We both knocked on their doors a good plenty. I was a sucker for those sales opportunities advertised on the back of comic books and sold lots of imprinted stationery to my sweet neighbors, enough to earn the Snoopy radio I'd coveted. Likely they locked the doors and turned off the lights when they saw us coming, but when we showed up with the mistletoe, they flung their doors and their change purses open wide with enthusiasm.

One year, the first person to answer the door bought the whole lot of our Christmas mistletoe. After that, under the guidance of my father who was a better snake man than businessman but who still recognized a flaw in our marketing, we doubled our prices. Eventually, we went all the way up to a dollar for our most expensive bundles; they still sold like hotcakes.

This all came back to me several years ago, when I found myself at a loss as to what to get my new coworkers when I began working at the Alabama State Department of Education. Good heavens, there were 20 of us, and we were exchanging small gifts for Christmas! Working in Montgomery, I didn't have time to trinket-shop.

Panicked, I called my husband Joe and asked if he'd fetch our late friend Kent and go out and shoot down some mistletoe for me to give my coworkers.

Here's the thing about love. Without question or hesitation, Joe called Kent, who stopped what he was doing immediately to join Joe in the mission. Now, I don't know where Joe and Kent went to shoot down the mistletoe. They wouldn't tell me. Times have changed, and there are laws about randomly shooting things out of trees in the neighbors' yards. All I know is I came home to a table-top full of mistletoe and two handsome men, grinning like mischievous elves, having a beer on the porch.

These days, the sight of mistletoe growing like a demon on the trees in the woods or hanging in a doorway like a relic of times gone by, makes me smile and warms my gullet.

Happy Solstice, one and all.

An Ant Atonement
December 28, 2024

Our little dog Peedie has to eat Hill's Science dog food. We're lucky that he likes it, and that he's so tiny he only eats a quarter of a cup a day. That stuff costs an arm and a leg, so last week, when ants invaded our bag of Hill's Science dog food, we went straight into beast mode and mercilessly attacked their trail with Windex. Horrified to find the entire bag of food completely inundated with ants, we threw the whole lot in the deep freeze to kill the pesky creatures inside, leaving us to pick out the dead ants before serving. Cussing ensues.

Typically, we remove uninvited critters from the house carefully, by hand, but ants in the Hill's was too much.

I feel bad about killing those ants. They're just trying to get along, strengthening the survival of their species like the rest of life on Earth, except maybe humans who seem to be hell-bent on destroying our ecosystem, but that's another story.

I keep thinking of E.O. Wilson, Alabama's own "father of biodiversity." Wilson's fascination with the little things, especially ants, propelled him toward his ultimate passion, that of striving for a sustainable existence for all life on our planet. And now I've gone and slaughtered hundreds and hundreds of little things.

To atone, I decided to read up on ants and will now share my worthy findings with the world, or at least with the readers of this column.

Our battle with the ants likely didn't do much harm to the ant population at large. There are about 20 quadrillion ants on the planet, with a biomass exceeding that of wild birds and mammals combined. Ants are found on every continent except Antarctica (ironically named), and on every region therein except Iceland, Greenland, and some islands. There are over 15,700 named species and subspecies of ants, ranging in size from teeny-tiny to one inch long.

Simply put, ants are plentiful.

As for the intrepid travelers in my kitchen, they are likely little back ants, odorous house ants, or Argentine ants. I'm guessing they're Argentine ants because they reportedly like dog and cat food. Perhaps they are little black ants, as they are little and black. I invite entomologists to chime in, especially myrmecologists, formerly known as formicologists, this shift due to reasons only etymologists can explain.

Whatever species they are, how the heck did the ants find that dog food amidst the world of smelly smells in our less-than-pristine kitchen? Ants send scouts out in search of food. These scouts have hundreds of nerve endings on their antennae, four to five times more than other insects, enabling them to find food from great distances. They then bring back a taste in one of their two stomachs for the colony's approval. If the food is deemed acceptable, the scouts return to the buffet, having memorized the trail with olfactory and visual clues, leaving behind a trail of pheromones to guide their colony mates. Hence the long, thin parade of ants to the dog food.

As you may remember from elementary school when you learned in fascination about ant colonies, then begged your parents for an ant farm, all the working ants are females. The males have only one job, to mate with the queen. Then they die. There is at least one ant species, however, with no males in the colonies.

Found in the Amazon, these exclusively female ant colonies reproduce by cloning. With all the ants having the exact same DNA, they are susceptible to pandemics and other threats. What could the evolutionary advantage be of this phenomenon, I wondered, the question I ask daily of the living world around me.

It turns out that the energy saved by not reproducing with males doubles the number of reproductive females born with each generation. Furthermore, with all that extra energy, these female ant colonies have cultivated some of the most elaborate and sophisticated fungi farms ever studied in ants, the world's first farmers.

In other news, a study published in January 2023 concluded that ants can sniff out cancer, joining dogs, mice, and even fruit flies as cancer detectors.

Baptiste Piqueret (Sorbonne Paris North University), lead author of this study, suggests that ants have an edge over dogs in their cancer sniffing prowess and are easier to train. Those wary of ants crawling around on their bodies needn't worry; this cancer detection is done with urine samples. The research is still far from clinical application, but it does hold promise.

Joe and I may have been able to save our dog food from the ant invasion, but in the long run, it may be the ants saving us. To quote E. O. Wilson, "If all mankind were to disappear, the world would regenerate back to the rich state of equilibrium that existed ten thousand years ago. If insects were to vanish, the environment would collapse into chaos."

Christmas for the Rescues
January 4, 2024

I awoke at 5:00 am like a kid on Christmas, raced through my coffee, and arrived at the barn in time to see the soft pinks and yellows of sunrise crest over the hills. Christmas for the Rescues, a fundraising event for Orr Farms Rescue & Sanctuary (orrfarms.org), was finally here. Chanoah, director of the sanctuary, met me at the barn, and we loaded my borrowed horses, Jasper and Shorty, into the trailer and got on our way.

Orr Farms was already abuzz with activity when we arrived. Volunteers were busy feeding the resident cows, goats, horses, and 17 dogs. Big metal trailers arrived carrying curious horses.

I left Jasper and Shorty, whispering promises into their anxiously perked ears that they would enjoy this special day, and headed to the event area where I was to assist with crafts and games. Unsure of my path, Christmas garlands and ornaments beckoned from the surrounding trees, assuring me of the right direction.

In no time, guests arrived. Small hands rolled pinecones in peanut butter and birdseed, making edible ornaments to hang in the forest, and decorated paper bags with leaves and twigs. Chilly guests warmed their gullets with hot cocoa and gathered around the fire to make s'mores.

Maple, a regal Missouri Fox Trotter rescued from a kill pen, tirelessly toted children on pony rides. Bright silver snowflakes sparkled against her gleaming black coat.

Hay bales and cozy blankets stoked anticipation for the upcoming hayride.

Our two granddaughters, Annabelle and Ruby, challenged the adults to pillowcase races. The adults lost both the races and their dignity.

And oh, my heart, Eddie and Eli, three-week-old baby goats, were available for cuddling, elicited squeals of adoration. I might've squealed.

Soon we heard jingle bells and the clip-clop of hooves coming up the path, and suddenly the event field was awash in a most dazzling collection of horses. Ribbons and sashes adorned their manes and tails; garlands, bells, and every sort of spangle gave these already gorgeous creatures an extra dose of pizzaz.

As the riders dismounted, I realized I was seeing saddles the likes of which I'd never seen before. They were traditional Mexican Charro saddles, ornately detailed, notable for their enormous saddle horns. I have much to learn about Mexican cowboys, or Vaqueros, many of whom were among the guests.

Once the horses were secured to trees, caterers set out lunch: steaming trays of mac-n-cheese, Brussels sprouts, roast beef, fried chicken, chocolate cake, rolls, and more. Hungry folks lined up for their comfort food.

I stood awestruck in the center of the event field. So many horses and humans, from infants to seniors, from so many cultures, speaking multiple languages and dressed in everything from Native American blankets to Wranglers and Stetsons.

There is something to be said for being surrounded by people coming together to celebrate not just horses but the worthy mission of an animal rescue and sanctuary. There is a poignance bordering on miraculous in standing with folks from such diverse cultures, in line for home cooked food on a cold December day, encircled by horses who pierce the ambiance with quiet nickerings and loud whinnies.

And there, last in the food line, stood Chanoah, a tiny dynamo, beaming with happiness and giddy exhaustion from the planning and coordinating of this event, as well as the daily operations of her rescue and sanctuary and her full-time job as a barn manager and horse trainer at G&L Farms, where I had the gracious fortune to meet her.

"As soon as we're done eating," she hollered. "we're going riding!"

Oh, my heavens, what a glorious ride. Chanoah led the way on her rescue horse Chief, followed by 27 horses and riders. After a few minutes of deciphering who could ride near whom to mitigate kicking, biting, and bolting, we were off through five miles of well-maintained trails, through creeks, across crests overlooking the elegant starkness of a winter woods, and into clearings where we could regroup, count heads, and return cowboy hats lost while loping up the hills.

My childhood horse-loving friend Lizzie and I rode side by side, along with the ghosts of our 12-year-old selves who'd dreamed of days just like this.

I stuck around till dark, attempting to help Chanoah with the tending of Jasper and Shorty who would have a sleepover in her barn before returning to their home in morning.

When I left, she still had horses to put away, cows and dogs to feed, and a passel of other chores. No doubt David, her husband, recuperating from surgery, was also ready for her company at the end of this day.

I hope Chanoah and David had time to appreciate the black and star-filled sky blanketing their precious slice of wonder in this world.

It is certain the animals she's rescued, living at her sanctuary, appreciate her.

Acknowledgements

My thanks go out to Brian Woodham for responding to my request to write for *The Auburn Villager*, and to Nikki and the late Don Eddins for giving my column space; to my readers, whose encouragement sparks me to keep on writing; to Joe and the rest of my family, for all those hours, especially on weekends, when I have my nose at the computer instead of being with you; to my friends who never complain at hearing all my ideas and who provide me with valuable feedback; and to all the wonders of the world for the endless awe.

My heart is full of gratitude.

About the Author

MARY DANSAK is a writer, a naturalist, an equestrian, and an animal lover. With a full career of science education behind her, she remains compelled to share her wonder and love of the natural world around her.

Mary lives in Alabama with her husband and three little dogs. Their three grown daughters and two granddaughters serve as a constant source of inspiration, as does their wooded yard, her childhood, and her frequent excursions throughout the state and beyond. As well as writing her weekly column, Mary works on a horse and cattle ranch as the family's writer and as an assistant to the equestrian care team.

Mary has co-authored four anthologies of essays, memoir, and fiction: *Be the Flame* (New Plains Press, 2011), *The Ploy of Cooking* (Upper Creek Publishing, 2012), *The Art of Wench Cooking* (Village Smith Publishers, 2018), and *The Mystic Memoir* (Villager Smith Press, 2024) with her writers' group, The Mystic Order of East Alabama Fiction Writers. She wrote several screenplays for the children's science television show, *Steve Trash Science*, and many of her works appear in a variety of print and online magazines and journals.

Made in the USA
Columbia, SC
16 February 2025

53888068R00136